How Science Wo

The new AQA GCSE Science specification incorporates two types of content:

- **Science Content** (example shown opposite). This includes all the scientific explanations and evidence that you need to be able to recall in your exams (objective tests or written exams). It is covered on pages 11–103 of the revision guide.
- **How Science Works** (example shown opposite). This is a set of key concepts, relevant to all areas of science. It is concerned with how scientific evidence is obtained and the effect it has on society. More specifically, it covers:
  - the relationship between scientific evidence and scientific explanations and theories
  - the practices and procedures used to collect scientific evidence
  - the reliability and validity of scientific evidence
  - the role of science in society and the impact it has on our lives
  - how decisions are made about the use of science and technology in different situations, and the factors affecting these decisions.

Because they are interlinked, your teacher may have taught the two types of content together in your science lessons. Likewise, the questions on your exam papers are likely to combine elements from both types of content. To answer them you will need to recall the relevant scientific facts and draw on your knowledge of how science works.

The key concepts from How Science Works are summarised in this section of the revision guide. You should be familiar with all of them, especially the practices and procedures used to collect scientific data (from all your practical investigations). Make sure you work through them all. Make a note of anything you are unsure about and then ask your teacher for clarification.

B1 Nerves and Hormones

B1.2 Nerves and hormones

Human bodies respond to internal and external changes. Plants respond to changes in external conditions. To understand this, you need to know:

- how the nervous system and hormones enable us to respond to external changes
- how nerves and hormones help us to control conditions inside our bodies
- how hormones are used in contraceptives and fertility treatments
- the effects of plant hormones.

Parts of the Nervous System

The nervous system consists of the **brain**, the **spinal cord**, the **neurones** and **receptors**. It allows organisms to react to their surroundings and to coordinate their behaviour. Information from receptors passes along **neurones** (nerve cells) to the brain. The brain coordinates the response, which is carried out by an **effector**. The effector may be a muscle, which responds by contracting, or a **gland**, which responds by releasing chemicals.

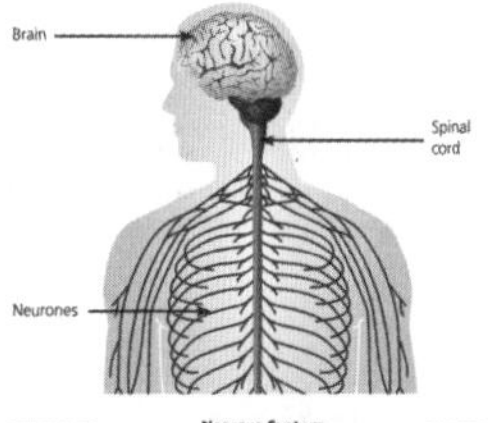

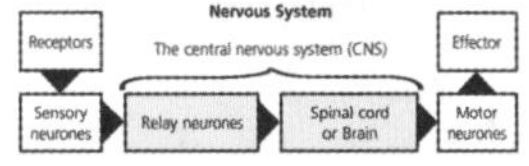

The Three Types of Neurone

Motor neurone – Impulse travels away from cell body

Sensory neurone – Impulse travels towards cell body

Relay neurone – Impulse travels first towards and then away from cell body

Neurones are specially adapted cells that can carry an electrical signal, e.g. a nerve impulse.

Neurones are elongated (stretched out) to make connections between parts of the body. They have branched endings that allow a single neurone to act on many muscle fibres. The cell body has many connections to allow communication with other neurones.

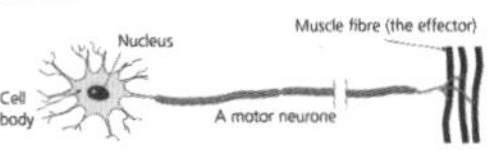

Connections Between Neurones

Neurones do not touch each other. There is a very small gap between them called a **synapse**.

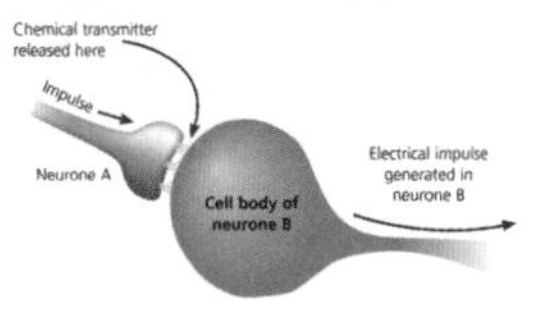

When an electrical impulse reaches the gap via neurone A, a chemical transmitter is released, which activates receptors on neurone B and causes an electrical impulse to be generated in neurone B. The chemical transmitter is then destroyed.

P1 How Science Works

You need to be able to evaluate the possible hazards associated with the use of different types of electromagnetic radiation.

Example

It's Good to Talk – or is it?

New findings raise concerns that mobile phones could cause cancer and other health problems.

Swedish scientists studying the effects of electromagnetic radiation on red blood cells have found that levels of radiation, equivalent to those emitted by mobile phones, have a significant effect on the attractive forces between cells.

Up until now, the conventional view has been that microwaves could only cause damage at a cellular level if they carried enough energy to break chemical bonds or 'fry' tissue. These new findings might suggest that mobile phones do in fact emit enough energy to affect the bonds.

Experts, however, have been quick to point out that the results were obtained through tests on small groups of cells and provide no real evidence of a danger to health.

There have been other suggestions in the past that mobile phones can cause brain tumours and Alzheimer's disease, but so far research has been inconclusive.

Mobile Phones are Safe

New study has found no link between mobile phones and cancer.

Scientists studying the possible links between mobile phones and brain tumours have reported today that they have found no correlation between using mobile phones and the risk of developing glioma – the most common type of brain tumour.

The study of 2 682 people across the UK looked at 966 individuals with diagnosed glioma and 1 716 individuals without the condition. It concluded that although a large percentage of the cancer sufferers reported their tumours to be on the side of the head where they held the phone, for regular mobile phone users, there was no increased risk of developing glioma.

Using Mobile Phones

Advantages	Disadvantages
• Easy, convenient method of communication, especially in a vulnerable situation – when car breaks down, alone at night, feel threatened, etc. • Can be used to access the Internet, take pictures, and watch television / video clips. • Easy way to keep in contact when away from home or abroad, e.g. text messages. • Many different tariffs and networks, which makes them affordable. • Can help in solving crime because mobile phones can be tracked.	• Some studies have linked mobile phone use with brain tumours and Alzheimer's disease. • The long-term effects of using mobile phones are not known – studies are still being carried out. • Increasingly, advertising is targeted at younger age groups who would be more vulnerable to any health implications.

How Science Works Overview

How Science Works

The AQA GCSE Science specification includes activities for each sub-section of science content. These require you to apply your knowledge of how science works and will help develop your skills when it comes to evaluating information, developing arguments and drawing conclusions.

These activities are dealt with on the How Science Works pages throughout the revision guide. Make sure you work through them all, because questions relating to the skills, ideas and issues covered on these pages could easily come up in the exam. Bear in mind that these pages are designed to provide a starting point from which you can begin to develop your own ideas and conclusions. They are not meant to be definitive or prescriptive.

Practical tips on how to evaluate information are included in this section, on page 9.

What is the Purpose of Science?

Science attempts to explain the world we live in. The role of a scientist is to collect evidence through investigations to:

- explain phenomena (i.e. explain how and why something happens)
- solve problems.

Scientific knowledge and understanding can lead to the development of new technologies (e.g. in medicine and industry), which have a huge impact on society and the environment.

Scientific Evidence

The purpose of evidence is to provide facts that answer a specific question, and therefore support or disprove an idea or theory.

In science, evidence is often based on data that has been collected by making observations and measurements.

A scientifically literate citizen should be equipped to question the evidence used in decision making. Evidence must therefore be approached with a critical eye. It is necessary to look closely at:

- how measurements have been made
- whether opinions drawn are based on valid and reproducible evidence rather than non-scientific ideas (prejudices and hearsay)
- whether the evidence is reproducible, i.e. it can be reproduced by others
- whether the evidence is valid, i.e. is reproducible, repeatable and answers the original question.

N.B. If data is not reproducible or repeatable, it cannot be valid.

To ensure scientific evidence is repeatable, reproducible and valid, scientists employ a range of ideas and practices that relate to the following:

1. **Observations** – how we observe the world.
2. **Investigations** – designing investigations so that patterns and relationships can be identified.
3. **Measurements** – making measurements by selecting and using instruments effectively.
4. **Presenting data** – presenting and representing data.
5. **Conclusions** – identifying patterns and relationships and making suitable conclusions.
5. **Evaluation** – considering the validity of data and appropriateness of methods used.

These six key ideas are covered in more detail on the following pages.

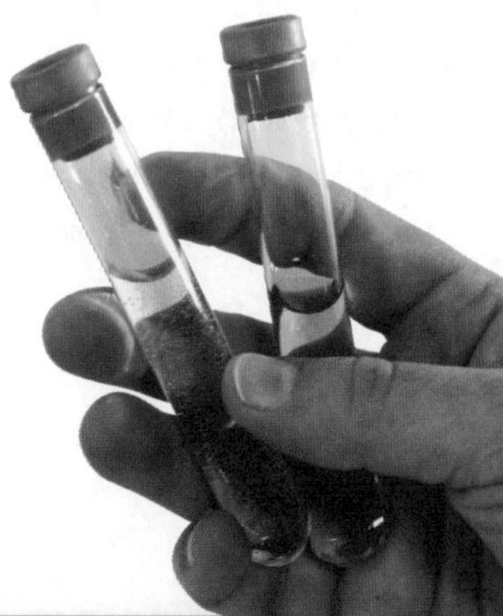

1 Observations

Most scientific investigations begin with an observation, i.e. a scientist observes an event or phenomenon and decides to find out more about how and why it happens.

The first step is to develop a **hypothesis**, i.e. to suggest an explanation for the phenomenon. Hypotheses normally propose a relationship between two or more **variables** (factors that can change). Hypotheses are based on careful observations and existing scientific knowledge, and often include a bit of creative thinking.

The hypothesis is used to make a prediction, which can be tested through scientific investigation. The data collected during the investigation might support the hypothesis, show it to be untrue, or lead to the development of a new hypothesis.

Example

A biologist **observes** that freshwater shrimp are only found in certain parts of a stream.

The biologist uses current scientific knowledge of freshwater shrimp behaviour and water flow to develop a **hypothesis**, which relates the distribution of shrimp (first variable) to the rate of water flow (second variable).

Based on this hypothesis, the biologist **predicts** that shrimp can only be found in areas of the stream where the flow rate is beneath a certain value.

The prediction is **investigated** through a survey that looks for the presence of shrimp in different parts of the stream, which represent a range of different flow rates.

The **data** shows that shrimp are only present in parts of the stream where the flow rate is below a certain value (i.e. the data supports the hypothesis). However, it also shows that shrimp are not *always* present in parts of the stream where the flow rate is below this value.

As a result, the biologist realises that there must be another factor affecting the distribution of shrimp. So, he **refines his hypothesis** to relate the distribution of shrimp (first variable) to the concentration of oxygen in the water (second variable) in parts of the stream where there is a slow flow rate.

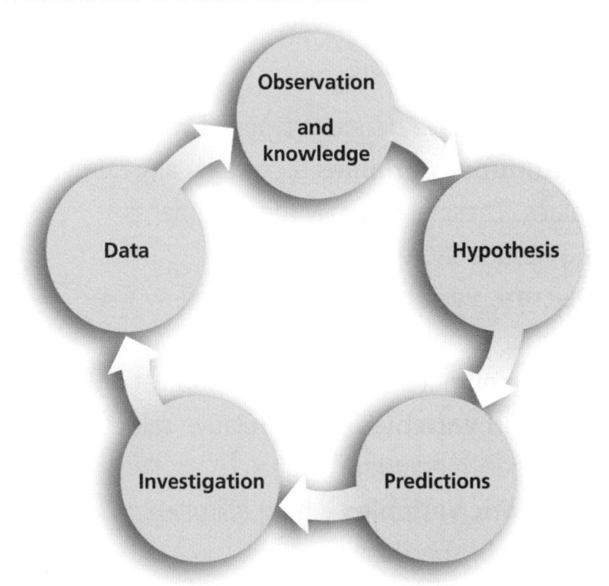

If new observations or data do not match existing explanations or theories (e.g. if unexpected behaviour is displayed) they need to be checked for reliability and validity.

In some cases it turns out that the new observations and data are valid, so existing theories and explanations have to be revised or amended. This is how scientific knowledge gradually grows and develops.

2 Investigations

An investigation involves collecting data to try to determine if there is a relationship between two variables. A variable is any factor that can take different values (i.e. any factor that can change). In an investigation there are three types of variables:

- **Independent variable**, which is changed or selected by the person carrying out the investigation. In the shrimp example on page 5, the independent variable is the flow rate of the water.
- **Dependent variable**, which is measured each time a change is made to the independent variable, to see if it also changes. In the shrimp example on page 5, the dependent variable is the number of shrimp.
- **Outside variables**, which are variables that could affect the dependent variable and must be controlled.

For a measurement to be valid it must measure only the appropriate variable.

Variables can have different types of values:

- **Continuous variables** – can take any numerical values. These are usually measurements, e.g. temperature or height.
- **Discrete variables** – can only take whole-number values. These are usually quantities, e.g. the number of shrimp in a population.
- **Ordered variables** – have relative values, e.g. small, medium or large.
- **Categoric variables** – have a limited number of specific values, e.g. the different breeds of dog: dalmatian, cocker spaniel, labrador, etc.

Numerical values tend to be more powerful and informative than ordered variables and categoric variables.

An investigation tries to establish if an observed link between two variables is one of the following:

- **Causal** – a change in one variable causes a change in the other, e.g. in a chemical reaction the rate of reaction (dependent variable) increases when the temperature of the reactants (independent variable) is increased.
- **Due to association** – the changes in the two variables are linked by a third variable. For example, a link between the change in pH of a stream (first variable) and a change in the number of different species found in the stream (second variable), may be the effect of a change in the concentration of atmospheric pollutants (third variable).
- **Due to 'chance occurrence', i.e. coincidence** – the change in the two variables is unrelated; it is coincidental. For example, in the 1940s the number of deaths due to lung cancer increased and the amount of tar being used in road construction increased. However, one increase *did not* cause the other increase.

2 Investigations (cont)

Fair Tests

A fair test is one in which the only factor that can affect the dependent variable is the independent variable. Any other variables (outside variables) that could influence the results are kept the same.

A fair test is much easier to achieve in the laboratory than in the field, where conditions (e.g. weather) cannot always be physically controlled. The impact of outside variables, like the weather, has to be reduced by ensuring that all measurements are affected by the variable in the same way. For example, if you were investigating the effect of different fertilisers on the growth of tomato plants, all the plants would need to be grown in a place where they were subject to the same weather conditions.

If a survey is used to collect data, the impact of outside variables can be reduced by ensuring that the individuals in the sample are closely matched. For example, if you were investigating the effect of smoking on life expectancy, the individuals in the sample would all need to have a similar diet and lifestyle to ensure that those variables did not affect the results.

Control groups are often used in biological research. For example, in some drugs trials a placebo (a dummy pill containing no medicine) is taken by one group of volunteers (the control group) and the drug is taken by another group. By comparing the two groups, scientists can establish if the drug (the independent variable) is the only variable affecting the volunteers and therefore whether or not it is a fair test.

Selecting Values of Variables

Care is needed in selecting the values of variables to be recorded in an investigation. For example, if you were investigating the effect of fertilisers on plant growth, you would need a range of fertiliser concentration that would give a measurable range of growth. Too narrow a range of concentration may fail to give any noticeable difference in growth. A trial run often helps to identify appropriate values to be recorded, such as the number of repeated readings needed and their range and interval.

Repeatability, Reproducibility and Validity

Repeatability measures how consistent results are in a single experiment. For example, a student measures the time taken for magnesium to react with hydrochloric acid. She repeats the test five times and finds that for each test the reaction takes 45 seconds. We can say that repeatability is high and the results are accurate.

In practice, it is unlikely that all five tests will give exactly the same result and therefore the mean (average) of a set of repeated measurements is often calculated to overcome small variations and get a best estimate of a true result.

An accurate measurement is one that is close to the true value. The purpose of an investigation will determine how accurate the data collected needs to be. For example, measurements of blood alcohol levels must be accurate enough to determine if a person is legally fit to drive.

Reproducibility measures the ability of an experiment to produce results that are the same each time it is carried out. For example, a whole class of students carry out the experiment above with magnesium and hydrochloric acid. The results can be said to be reproducible, and therefore reliable, if all students' results are very close to each other.

Validity questions if the results obtained can be used to prove or disprove the original hypothesis. It must consider the design of the experiment, the extent to which variables have been controlled and the reliability of results. The data collected must be precise enough (i.e. to an appropriate number of decimal places) to form a valid conclusion.

How Science Works Overview

❸ Measurements

When making measurements, errors may occur that affect the repeatability, reproducibility and validity of the results. These may be due to:

- **Variables that have not been controlled** – these may be variables that are beyond control.
- **Human error** – when making measurements, random errors can occur due to a lapse in concentration. Random errors can also result from inconsistent application of a technique. Systematic (repeated) errors can result from an instrument not being calibrated correctly or repeatedly being misused, or from consistent misapplication of a technique.
- **The resolution of the instruments used** – the resolution of an instrument is determined by the smallest change in value it can detect. For example, the resolution of bathroom scales is insufficient to detect the changes in mass of a small baby, whereas the scales used by a midwife have a higher resolution. It is therefore important to select instruments with an appropriate resolution for the task.

There will always be some variation in the actual value of a variable no matter how hard we try to repeat an event. For example, if the same model parachute was dropped twice from the same height, in the same laboratory and how long it took to fall to the ground was timed, it is very unlikely both drops would take exactly the same time. However, the two results should be close to each other. Any anomalous values should be examined to try to identify the cause. If the result is a product of a poor measurement, it should be ignored.

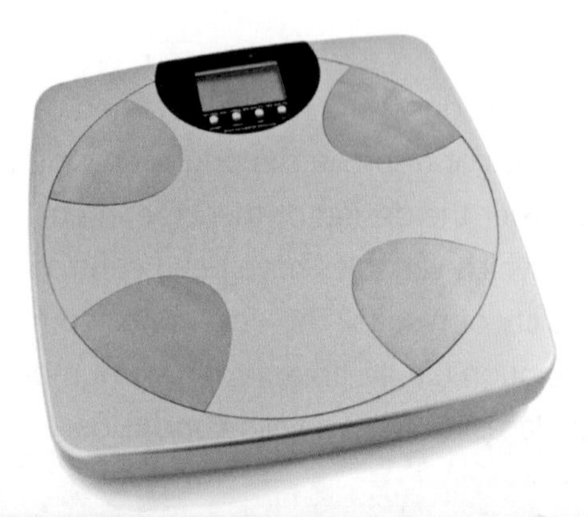

❹ Presenting Data

When presenting data, two terms are frequently used:

- the **mean** (or average) – this is the sum of all the measurements divided by the number of measurements taken
- the **range** – this refers to the maximum and minimum values and the difference between them.

To explain the relationship between two or more variables, data can be presented in such a way that the pattern is more evident. The type of presentation used will depend on the type of variable represented.

Tables are an effective way to display data, but are limited in how much they can tell us about the design of an investigation.

Height (cm)	127	165	149	147	155	161	154	138	145
Shoe size	5	8	5	6	5	5	6	4	5

Bar charts are used to display data when one of the variables is categoric. They can also be used when one of the variables is discrete.

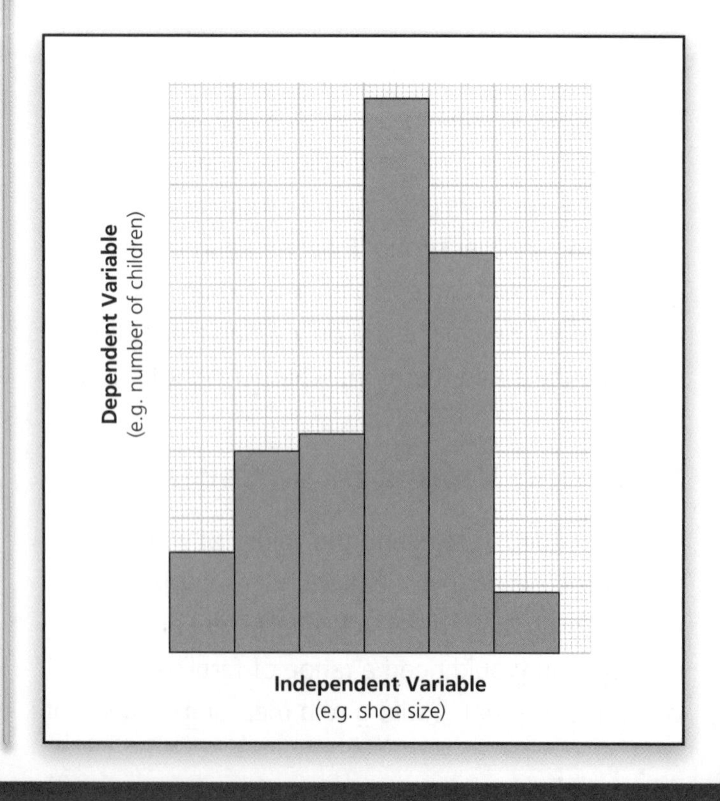

Scattergrams can be used to show an association between two variables. This can be made clearer by including a line of best fit.

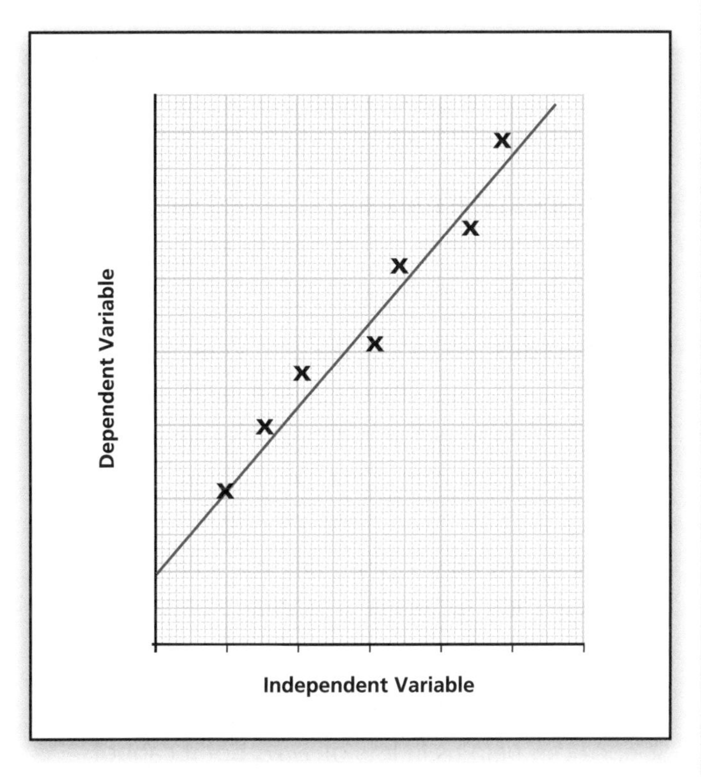

Line graphs are used when both the dependent and independent variables are continuous.

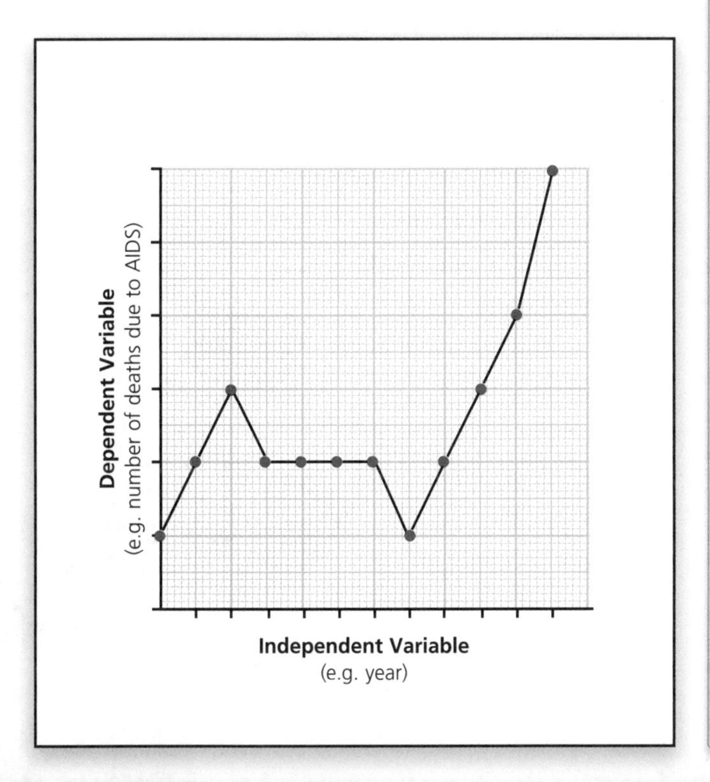

5 Conclusions

The patterns and relationships observed in the data represent what has happened in the investigation. However, it is necessary to look at the patterns and trends between the variables (bearing in mind limitations of the data), in order to draw conclusions.

Conclusions should:

- describe the patterns and relationships between variables
- take all the data into account
- make direct reference to the original hypothesis / prediction.

Conclusions should not:

- be influenced by anything other than the data collected
- disregard any data (other than anomalous values)
- include any speculation.

6 Evaluation

In evaluating a whole investigation, the repeatability, reproducibility and validity of the data obtained must be considered. This should take into account the original purpose of the investigation and the appropriateness of methods and techniques used in providing data to answer the original question.

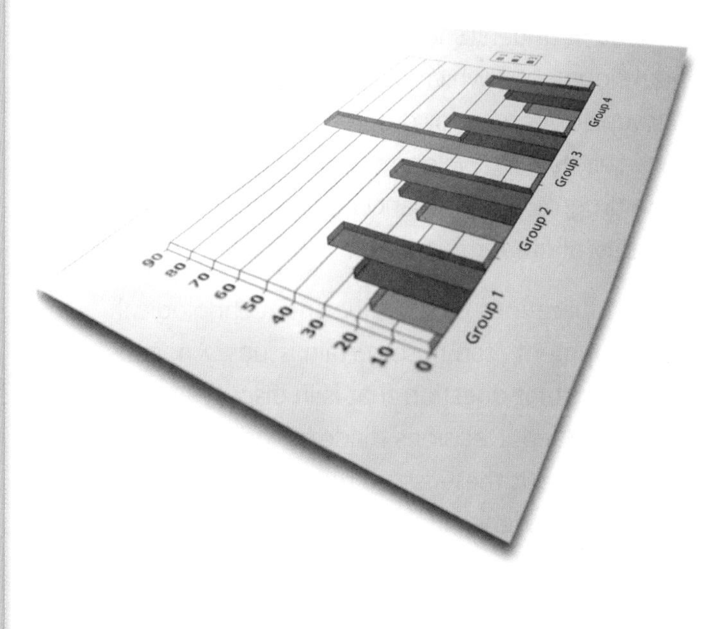

How Science Works Overview

Societal Aspects of Scientific Evidence

Judgements and decisions relating to social-scientific issues may not be based on evidence alone. Sometimes other factors may have an influencing role. For example:

- **Bias** – evidence must be scrutinised for any potential bias on the part of the experimenter. Biased information might include incomplete evidence or may try to influence how you interpret the evidence. For example, this could happen if funding for the investigation came from a party with a vested interest, e.g. a drug company wanting to highlight the benefits of their new drug but downplay the side effects in order to increase sales.
- **Weight of evidence** – evidence can be given undue weight or be dismissed too lightly due to:
  - political significance – if the consequences of the evidence were likely to provoke public or political unrest or disquiet, the evidence may be downplayed
  - status of the experimenter – evidence is likely to be given more weight if it comes from someone with academic or professional status or who is considered to be an expert in that particular field.

Science and Society

Scientific understanding can lead to technological developments, which can be exploited by different groups of people for different reasons. For example, the successful development of a new drug benefits the drug company financially and improves the quality of life for patients.

The applications of scientific and technological developments can raise certain issues. An 'issue' is an important question that is in dispute and needs to be settled. Decisions made by individuals and society about these issues may not be based on scientific evidence alone.

Social issues are concerned with the impact on the human population of a community, city, country, or even the world.

Economic issues are concerned with money and related factors like employment and the distribution of resources. There is often an overlap between social and economic issues.

Environmental issues are concerned with the impact on the planet, i.e. its natural ecosystems and resources.

Ethical issues are concerned with what is morally right and wrong, i.e. they require a value judgement to be made about what is acceptable. As society is underpinned by a common belief system, there are certain actions that can never be justified. However, because the views of individuals are influenced by lots of different factors (e.g. faith and personal experience) there are also lots of grey areas.

Limitations of Scientific Evidence

Science can help us in lots of ways but it cannot supply all the answers. We are still finding out about things and developing our scientific knowledge. There are some questions that we cannot answer, maybe because we do not have enough reproducible, repeatable and valid evidence.

There are some questions that science cannot answer at all. These tend to be questions relating to ethical issues, where beliefs and opinions are important, or to situations where we cannot collect reliable and valid scientific evidence. In other words, science can often tell us *whether* something can be done or not and *how* it can be done, but it cannot tell us if it *should* be done.

B1.1 Keeping healthy

A healthy diet and regular exercise are required to keep our bodies healthy. To understand this, you need to know:

- what a healthy diet consists of and the consequences of a poor diet
- how exercise affects the body
- how the body defends itself against microbes
- how vaccination can be used to prevent infection.

Healthy Diets

A healthy diet contains the correct balance of the different foods that our body needs, i.e. carbohydrates, fats, proteins, fibre, vitamins, minerals, water – and the right amount of energy.

Carbohydrates, proteins and fats can all be used by the body to release energy and to build cells. Minerals and vitamins are needed in small amounts for healthy functioning of the body.

A person is **malnourished** if their diet is not balanced. A poor diet can lead to:

- a person being too fat (overweight) or too thin (underweight)
- **deficiency diseases** and conditions such as Type 2 diabetes.

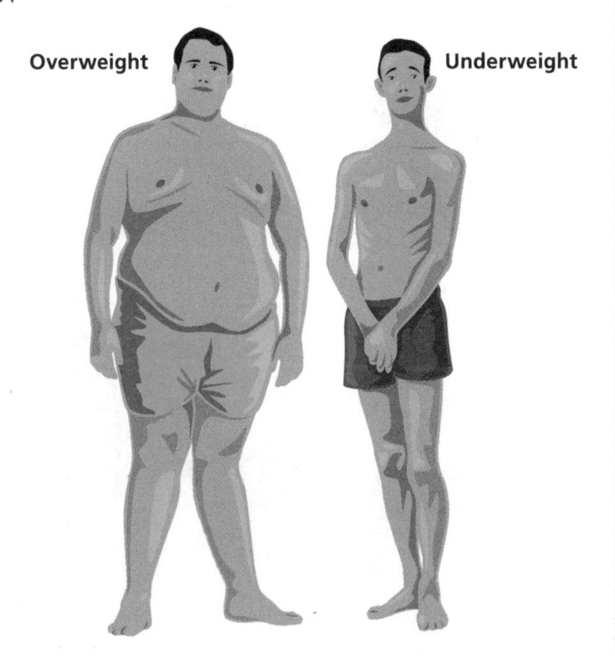

Exercise

Exercise increases the amount of energy expended by the body.

A person will lose mass if the energy content of food taken in is less than the amount of energy expended by the body. This is the basis for most slimming programmes.

People who exercise regularly are usually healthier than those who take little exercise.

Metabolic Rate

The **metabolic rate** is the rate at which all the chemical reactions in the cells of the body are carried out. This rate varies with:

- the amount of activity you do – your metabolic rate increases with the amount of exercise you do and it stays high for some time after you have finished exercising
- the proportion of fat to muscle in your body.

Inherited Factors Affect Health

Our metabolic rate may also be affected by inherited factors. Cholesterol is made naturally in the liver. Diet and inherited factors affect how much cholesterol the liver makes.

You need to be able to evaluate information about the effect of food on health and about the effect of lifestyle on development of disease.

Example

There is much debate over what constitutes a healthy lifestyle, and what's healthy for one person might be completely inappropriate for another. However, diet has a major influence on health and there are a few guidelines that will benefit everyone.

Evidence suggests that the most healthful diets do not include large amounts of animal products, which are often high in saturated fats and cholesterol. Instead, plenty of fruit and vegetables are recommended together with wholemeal foods, which are high in fibre.

'Are vegetarian diets healthy?' is a question often asked. Studies have shown that vegetarians often have lower incidence of coronary artery disease, hypertension, obesity and some forms of cancer. This is because vegetarian diets tend to be higher in polyunsaturated fats (usually from plants) and lower in saturated fat (which comes mostly from animals).

Watch the amount of food you eat as well. Between 20% and 30% of the UK population is thought to be overweight or obese. Being obese or overweight substantially increases the risk of developing numerous conditions, including heart disease, high blood pressure, Type 2 diabetes, stroke, osteoarthritis and several types of cancer.

The average person consumes six to eight times more salt than is needed each day. So, reducing this intake is important for anyone who wants to eat a healthy diet. A high-salt diet is linked to high blood pressure, which can lead to heart disease, stroke and kidney failure. There is often a lot of hidden salt in processed food, therefore reducing the amount of processed food we eat will help us towards this goal.

And finally, exercise benefits everyone, although it is important to have a programme tailored to the individual's capabilities and needs. The physical benefits of exercise include joint lubrication for those with arthritis, bone rebuilding for those with osteoporosis, and lowering of cholesterol. Regular exercise can also help control weight.

For a healthy lifestyle we should:

- eat less animal products
- eat more fruit and vegetables and wholemeal foods
- eat less salt
- exercise regularly.

This will reduce the risk of developing diseases later in life such as:

- heart disease and high blood pressure
- stroke
- Type 2 diabetes
- osteoarthritis.

More of:

Less of:

You need to be able to evaluate claims made by a slimming programme.

Example

Slimmer April 2010

Advertorial

The all-new low-carb diet

It's easy to lose weight quickly on this new slimming programme. All you have to do is limit your carbohydrate intake.

You can consume a very small amount of carbohydrates from vegetables such as broccoli, but you are not allowed any carbohydrates from bread, pasta or fruit.

- Eat unlimited quantities of protein and fats (e.g. meat, cheese, eggs)!
- No need to make yourself eat fruit!
- No need for exercise!

How does it work?

Managing the intake of carbohydrates is certain to lead to weight loss because without carbohydrates to provide energy, the body starts using up stored fat as the main source of energy.

Benefits	Problems
• Low-carbohydrate diets can help you to lose weight very quickly in the short term. • Low-carbohydrate diets can seem like an easy way to lose weight. • You can still eat many of the foods that would be cut out of a conventional slimmer's diet, e.g. sausages, cheese.	• Such a drastic change in diet can have a negative effect on the body. • Exercise is not encouraged by this slimming programme, but it is required for a person to be healthy – especially when they are consuming so much fat. • The body needs a healthy balance of foods to maintain good health: low-carbohydrate diets are not balanced and not good in the long term. • The body misses out on essential nutrients found in fruit, vegetables and grains. • Low-carbohydrate diets do not allow sufficient consumption of fruit and vegetables to meet the recommended daily allowance for carbohydrates. • Long-term high intake of protein puts a strain on the kidneys. • Lack of energy (energy should come from carbohydrates). • In the absence of carbohydrates, protein (muscle) stores get used for energy, as well as fat stores.

Bacteria and Viruses

Microorganisms that cause **infectious** diseases are called **pathogens**. Bacteria and viruses are the two main types of pathogen that may affect health.

Bacteria	Viruses
Very small	Even smaller than bacteria
Reproduce very quickly	Reproduce very quickly once inside living cells, which are then damaged
Can produce toxins (poisons) that make us feel ill	Can produce toxins (poisons) that make us feel ill
Responsible for illnesses like tetanus, cholera, tuberculosis	Responsible for illnesses such as colds, flu, measles, polio

Defence Against Pathogens

White blood cells form part of the body's **immune system**. They help to fight infection by:
- ingesting pathogens
- producing antitoxins to neutralise **toxins** produced by some pathogens
- producing **antibodies** to destroy particular pathogens. This leads to immunity from that pathogen.

Treatment of Disease

The symptoms of disease are often alleviated using painkillers. You will be familiar with the mild versions of these, e.g. aspirin. Although painkillers are useful, they do not kill pathogens.

Antibiotics

Antibiotics (e.g. **penicillin**) are often used to kill bacteria inside the body. It is important that specific bacteria are treated with specific antibiotics. However, antibiotics cannot be used to kill viruses, which live and reproduce inside cells. It is difficult to develop drugs that destroy viruses without damaging the body's tissues.

Overuse of Antibiotics

The use of antibiotics has greatly reduced deaths from infectious bacterial diseases. However, many strains of bacteria, including **MRSA** (Methicillin-resistant *Staphylococcus aureus*) have developed resistance to antibiotics through overuse and inappropriate use of antibiotics.

Resistance happens when the bacteria mutate to produce a new resistant strain. Scientists, therefore, continually need to develop new antibiotics. To prevent further resistance it is important to avoid overuse of antibiotics.

HT Some individual bacteria in a particular strain may have natural resistance to an antibiotic. When bacteria of the non-resistant strain are killed by the antibiotic, the resistant pathogens survive and reproduce. Thus, the population of the resistant strain increases.

Nowadays, in order to slow down the rate of development of resistant strains, antibiotics are not used to treat non-serious infections, e.g. sore throats.

Vaccination

A person can acquire **immunity** to a particular disease by being vaccinated (immunised).

1. An inactive / dead pathogen (a **vaccine**) is injected into the body.
2. White blood cells produce antibodies to destroy the pathogen.
3. The body now has an acquired immunity to this particular pathogen because the white blood cells are sensitised to it and will respond to any future infection by producing antibodies quickly. An example is the MMR vaccine used to protect children against measles, mumps and rubella.

If a large proportion of the population are immune to a pathogen, the spread of the pathogen is very much reduced.

Preparing a Culture Medium

Microorganisms can be grown in a **culture medium** containing various nutrients that the particular microorganism may need. These may include:

- carbohydrates – as an energy source
- mineral **ions**
- vitamins
- proteins.

Agar is most commonly used as the growth medium. This is a soft, jelly-like substance that melts easily and re-solidifies at around 50°C. The nutrients mentioned above are added to the agar to provide ideal growing conditions for cultures.

Preparing Uncontaminated Cultures

Uncontaminated cultures of microorganisms are required for investigating the action of antibiotics and disinfectants.

If the cultures we want to investigate become contaminated by unwanted microorganisms, these 'rogue' microorganisms may produce undesirable substances that can prove harmful.

It is only safe to use microorganisms if we have a pure culture of one particular species of microorganism.

To make useful products, uncontaminated cultures of microorganisms are prepared using the following procedures.

1. **Sterilisation of petri dishes and culture medium**
   Both **petri dishes** and the culture medium are **sterilised** using an autoclave – this is a pressure cooker that exposes the dishes and the agar to high temperature and high pressure to kill microorganisms.

2. **Sterilisation of inoculating loops**
   Inoculating loops are normally made of nichrome wire inserted into a wooden handle. They should be picked up like a pen and the loop and half the wire should be heated to red heat in a Bunsen flame, before being left to cool for five seconds. They are then sterile and can be used safely to transfer microorganisms to the culture medium.

   N.B. *Do not blow on the loop or wave it around to cool it as it will pick up more microorganisms.*

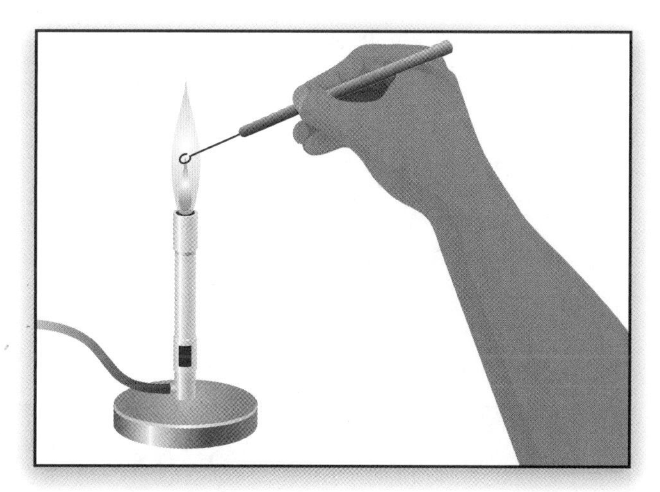

3. **Sealing the petri dish**
   After the agar has been poured in and allowed to cool, the petri dish should be sealed with tape (to prevent microorganisms in the air from entering) and clearly labelled on the base. It should be stored upside down so that condensation forms in the lid.

   In schools and colleges, cultures should only be **incubated** at a maximum of 25°C to prevent the growth of pathogens that grow at body temperature (37°C) and are potentially harmful to humans. In industry, higher temperatures can be used to produce more rapid growth.

You need to be able to evaluate the advantages and disadvantages of being vaccinated against a particular disease.

The MMR Vaccine

In Japan, scientists undertook a study of over 30 000 children born in the same city to compare the number of cases of autism in children who had been given the MMR triple vaccine compared with those who had not.

They counted the number of children who had been diagnosed with autism by the age of seven.

For children born in 1990, which was one year after the MMR vaccine was made compulsory, they found there were 86 cases of autism per 10 000 children. In 1991 the figure was 56 per 10 000 cases. The MMR vaccine was withdrawn in 1993 and children were given three separate injections. For children born a year after withdrawal of the vaccine, the number of cases of autism was 161 per 10 000 children.

Because the number of cases of autism rose after the MMR vaccine was withdrawn, the scientists concluded that the MMR vaccine was not responsible.

Benefits of MMR vaccination

- The MMR vaccine protects children against three potentially fatal diseases – measles, mumps and rubella.
- The widespread use of the vaccine prevents epidemics of any of the diseases.
- The triple vaccine means children only have to have one injection.

Problems with vaccination against MMR

- Smaller studies have suggested a link between the MMR vaccine and autism in children.
- Some larger studies do not rule out that MMR may trigger autism in a small number of children.

You need to be able to relate the contribution of Semmelweiss in controlling infection to solving modern problems with the spread of infection in hospitals.

Ignaz Semmelweiss (1818–1865) was a Hungarian doctor. He is remembered for his work in local hospitals where he reduced patient deaths on his wards from 12% to 1% by insisting that doctors washed their hands after surgery and before visiting another patient. He had recognised that 'something' on surgeons' and doctors' hands were infectious and contagious and were responsible for many patients' deaths.

The work of Semmelweiss, and subsequent scientists, has led to the creation of many modern regulations that help to maintain hygiene standards in modern hospitals and reduce the chance of infections being spread.

Health and Safety Regulations

- All staff must wash their hands thoroughly before and after having contact with each patient.
- All patients must wash their hands thoroughly.
- All surgical instruments must be sterilised before use.
- All hospital wards must be cleaned regularly with antibacterial cleaner.
- Doctors and surgeons must wear disposable face masks, gowns and gloves.
- All spillages of body fluids must be cleared up immediately.
- All patients with infectious diseases must be isolated to prevent disease spreading.

You need to be able to explain how the treatment of disease has changed as a result of increased understanding of the action of antibiotics and immunity...

... and evaluate the consequences of mutations of bacteria and viruses in relation to epidemics and pandemics, e.g. bird influenza.

Example

Genetically Modified Vaccine may be Answer to Bird Flu

Currently, scientists are trying to prevent an epidemic (large-scale infection) of bird influenza becoming pandemic (a world-wide infection) by developing new vaccinations. There are fears that the bird flu virus could mutate into a strain that can be transmitted between humans. US scientists are currently working on a genetically engineered vaccine that has been found to protect mice from the strains of bird flu that recently killed people in Asia and Europe. The vaccine, which was created from a genetically modified common cold virus, has been shown to stimulate the white blood cells to produce specific antibodies that may fight a number of strains of the bird flu virus.

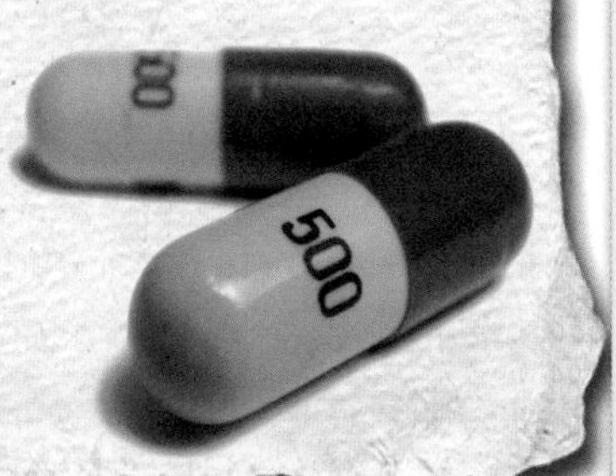

The treatment of disease changed drastically with the understanding of antibiotics and immunity.

Antibiotics are medicines that kill bacteria, so many bacterial infections and diseases, e.g. bronchitis, can be treated quite simply with a course of antibiotics. Patients suffering from illnesses that could once have led to death, now have a good chance of recovery through the use of antibiotics.

People can acquire immunity to many diseases through vaccination, which means many once common diseases can now be prevented. Today, most young children are given vaccinations to immunise them against diphtheria, whooping cough, polio, meningitis and tuberculosis. Most are also given the MMR vaccine to immunise them against measles, mumps and rubella.

Vaccines are also available for tourists visiting places such as some parts of Africa and Asia, to protect them from diseases such as rabies and yellow fever. These diseases and illnesses could be fatal without modern medicine.

So what are the new problems facing today's scientists researching the prevention and cure of disease?

1. Some strains of bacteria have developed resistance to antibiotics as a result of natural selection, e.g. MRSA. This results in the need for new antibiotics to be developed.
2. Some viruses mutate rapidly (change their form) so that existing vaccines are no longer effective. The influenza (flu) virus can change rapidly, so a vaccine which combated its effect on the body one year will no longer be effective the following year. The new strain will spread rapidly because people are not immune to it and there is no effective treatment. This means scientists are always having to develop new vaccines and find new ways to protect against these new strains.

B1.2 Nerves and hormones

Human bodies respond to internal and external changes. Plants respond to changes in external conditions. To understand this, you need to know:

- how the nervous system and hormones enable us to respond to external changes
- how nerves and hormones help us to control conditions inside our bodies
- how hormones are used in contraceptives and fertility treatments
- the effects of plant hormones.

Parts of the Nervous System

The nervous system consists of the **brain**, the **spinal cord**, the **neurones** and **receptors**. It allows organisms to react to their surroundings and to coordinate their behaviour. Information from receptors passes along **neurones** (nerve cells) to the brain. The brain coordinates the response, which is carried out by an **effector**. The effector may be a muscle, which responds by contracting, or a **gland**, which responds by releasing chemicals.

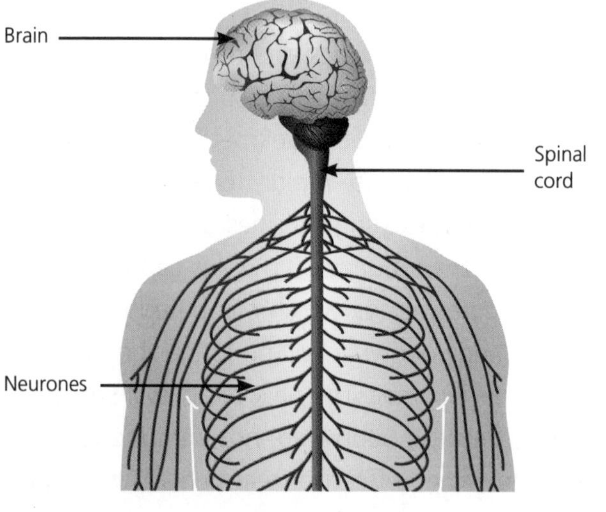

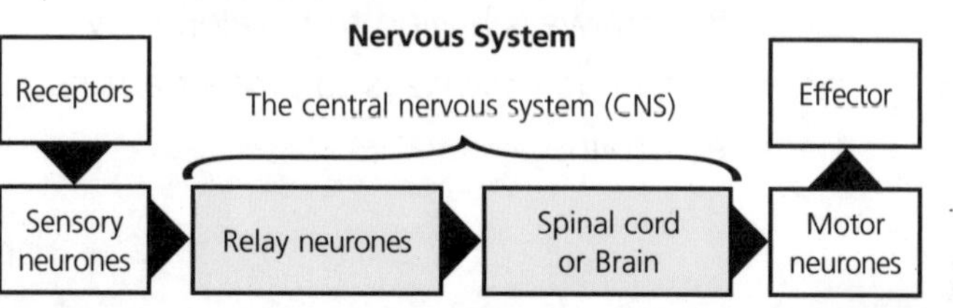

The Three Types of Neurone

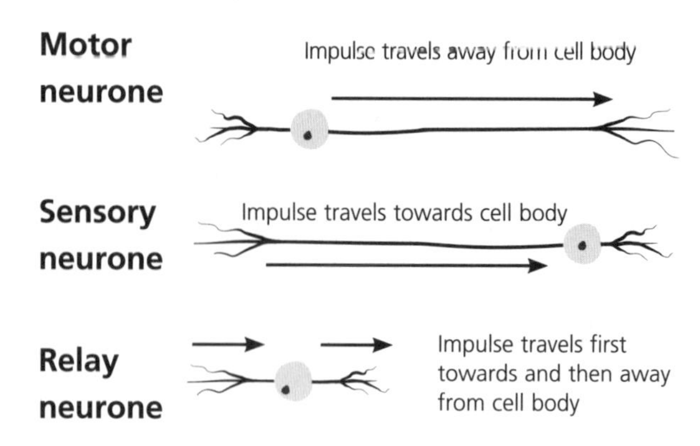

Neurones are specially adapted cells that can carry an electrical signal, e.g. a nerve impulse.

Neurones are elongated (stretched out) to make connections between parts of the body. They have branched endings that allow a single neurone to act on many muscle fibres. The cell body has many connections to allow communication with other neurones.

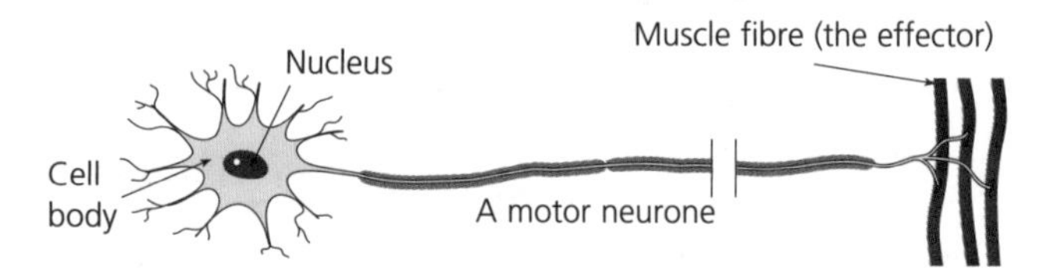

Connections Between Neurones

Neurones do not touch each other. There is a very small gap between them called a **synapse**.

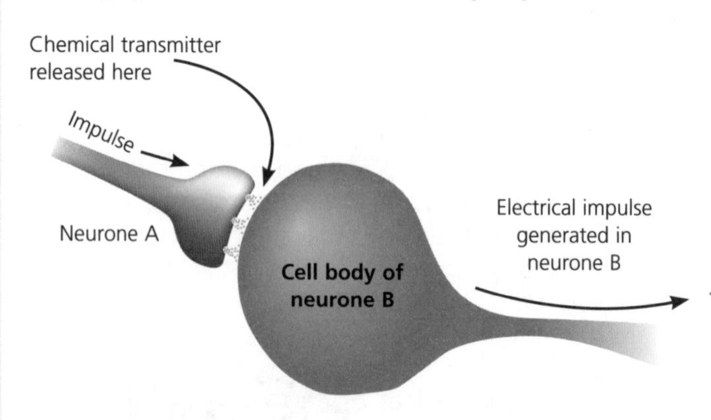

When an electrical impulse reaches the gap via neurone A, a chemical transmitter is released, which activates receptors on neurone B and causes an electrical impulse to be generated in neurone B. The chemical transmitter is then destroyed.

Types of Receptor

Receptors detect (respond to) stimuli that include:

- **Light –** receptors in the eyes.
- **Sound –** receptors in the ears.
- **Change of position –** receptors in the ears (balance).
- **Taste –** receptors on the tongue are sensitive to chemicals.
- **Smell –** receptors in the nose are sensitive to chemicals.
- **Touch, pressure, pain and temperature –** receptors in the skin.

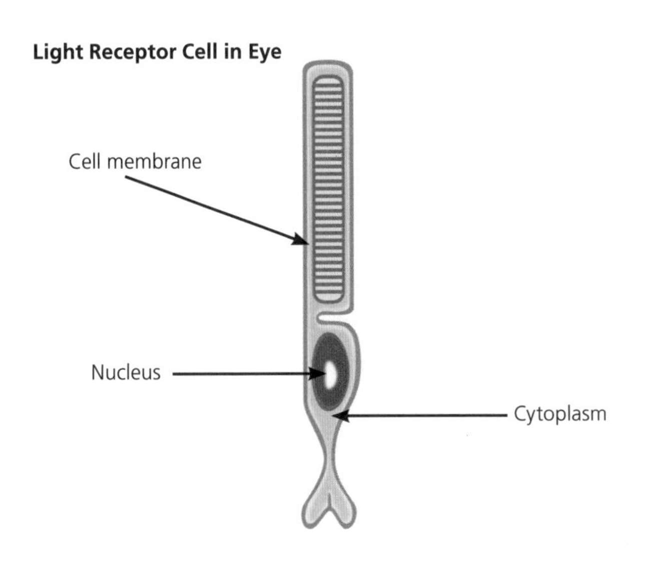

Reflex Action

Sometimes, conscious action is too slow to prevent harm to the body, e.g. removing your hand from a hot plate. **Reflex action** speeds up the response time by missing out the brain completely. The spinal cord acts as the coordinator and passes impulses directly from a sensory neurone to a motor neurone via a relay neurone. This by-passes the brain. Reflex actions are therefore automatic and quick.

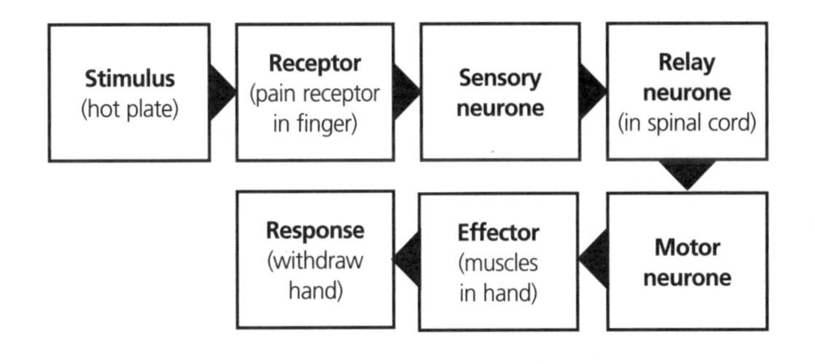

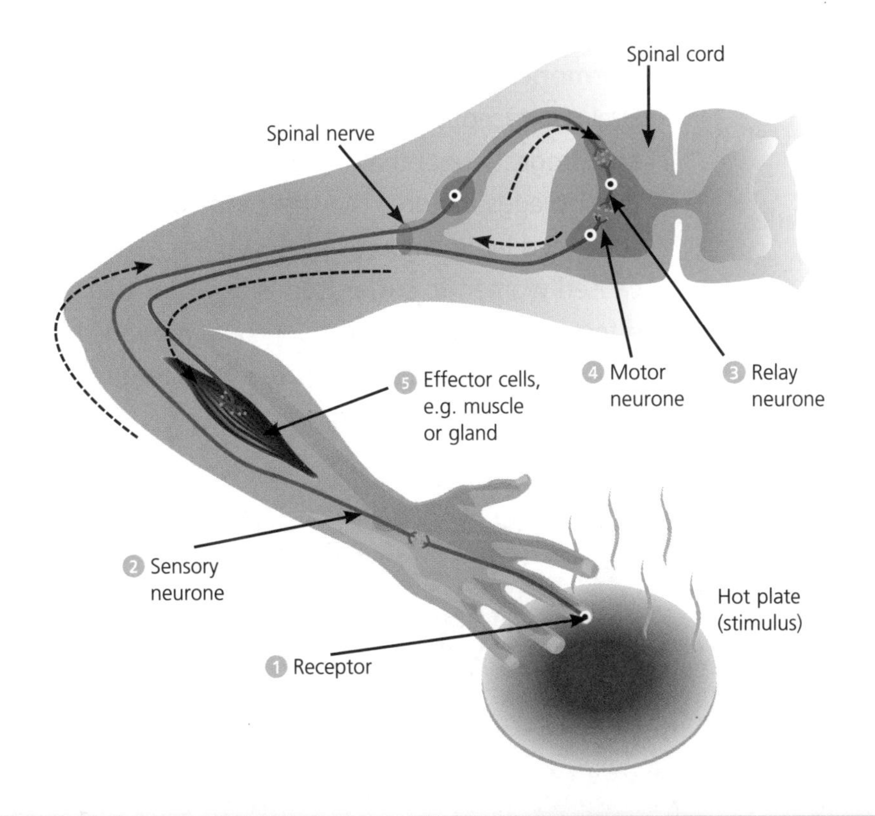

1. A receptor is stimulated by the hot plate (stimulus)…
2. …causing impulses to pass along a sensory neurone into the spinal cord.
3. The sensory neurone synapses with a relay neurone, by-passing the brain. A chemical is released at the synapse, which causes an impulse to be sent along the relay neurone.
4. A chemical is released when the relay neurone synapses with the motor neurone. This causes an impulse to be sent down the motor neurone…
5. …to the muscles (effectors) causing them to contract to move hand from hotplate.

Controlling Internal Conditions

Humans need to keep their internal environment relatively constant. Body temperature, and the levels of water, salts (ions) and blood sugar need to be carefully controlled.

How Conditions are Controlled

Many processes within the body (including control of some internal conditions) are coordinated by **hormones**. Hormones are chemicals, produced by glands, which are transported to their target organs by the bloodstream.

Hormones and Fertility

Hormones regulate the functions of many organs and cells.

A woman naturally produces hormones that cause:

- the monthly release of an egg from her ovaries
- changes in the thickness of the lining of her womb.

These hormones are produced by the **pituitary gland** and the **ovaries**.

Natural Control of Fertility

A woman's fertility is naturally controlled by the body.

1. **Follicle stimulating hormone** (**FSH**) is secreted by the pituitary gland. FSH causes eggs to mature in the ovaries and it causes the ovaries to produce hormones, including oestrogen.
2. **Oestrogen**, produced in the ovaries, inhibits further production of FSH and causes the production of **luteinising hormone** (LH).
3. **LH**, also from the pituitary gland, stimulates the release of an egg from the ovaries in the middle of the **menstrual** cycle.

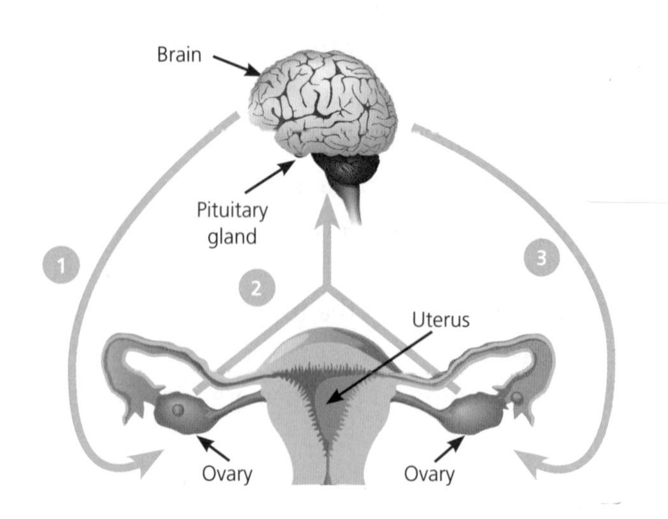

Artificial Control of Fertility

Hormones can be used to control fertility.

Hormones increasing fertility

LH and FSH can be given in a fertility drug to a woman whose own levels of FSH are too low to stimulate eggs to mature. This happens in *in vitro* fertilisation (IVF) treatment where several eggs are stimulated to mature. The eggs are collected from the mother, are fertilised by the father's sperm and then develop into **embryos**. At this stage, when they are tiny balls of cells, one or two embryos are inserted into the mother's uterus (womb).

Hormones reducing fertility

Oral contraceptives contain hormones (oestrogen and progesterone) that inhibit FSH production so that no eggs mature. The first birth control pills contained large amounts of oestrogen, which caused women to suffer from significant side effects. Birth control pills now contain a much lower dose of oestrogen or are progesterone only, resulting in fewer side effects.

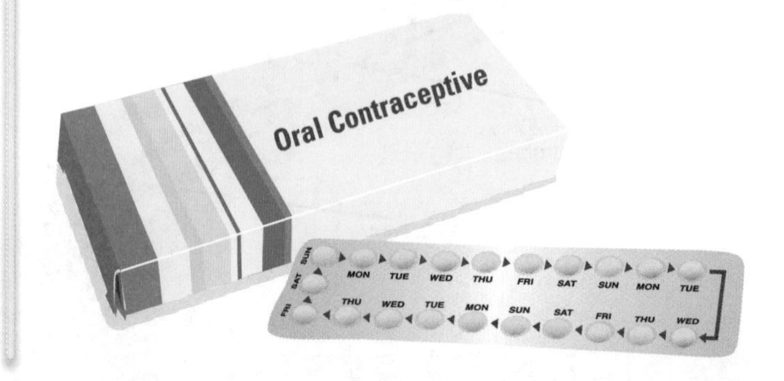

Hormones and Control of Plants

Plants are sensitive to:

- **light** – their shoots will grow towards light
- **moisture** – their roots grow towards moisture
- **gravity** – roots grow in the direction of gravity whilst shoots grow against the force of gravity.

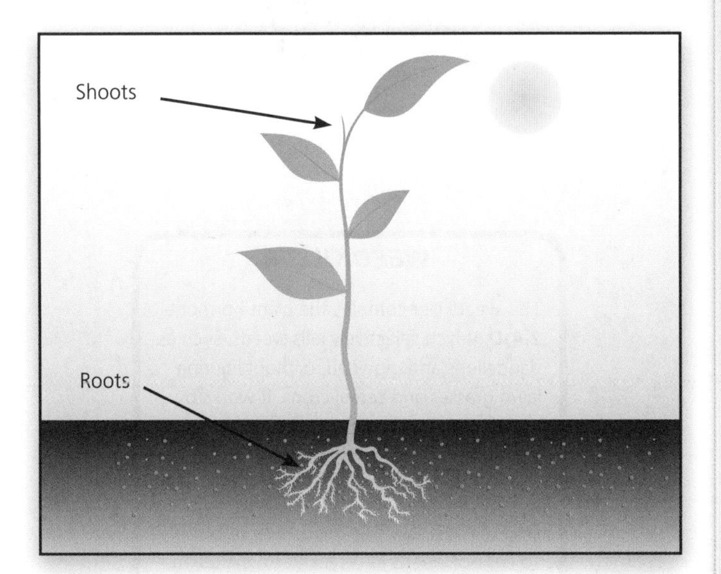

Auxins

Auxins are hormones that control the growth of plants at the tips of roots and shoots. Auxins promote growth of shoots, but inhibit growth in roots.

When a shoot tip is exposed to light, more auxin is produced on the side of the shoot that is in the shade. This causes the shoot to grow faster on the shaded side and the plant will bend towards the light. This is known as **phototropism**.

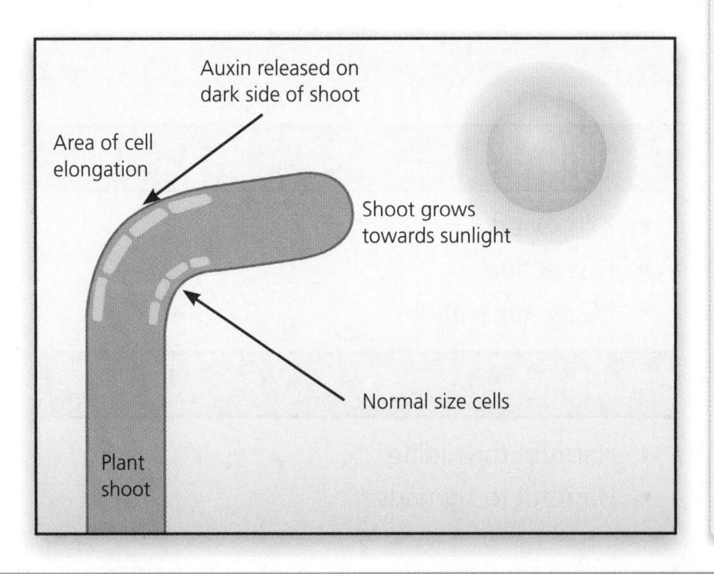

If a plant root is growing laterally (horizontally), gravity will cause more auxin to be produced on the lower side. The extra auxin inhibits growth causing the root to bend downwards. This is called **gravitropism** or **geotropism**.

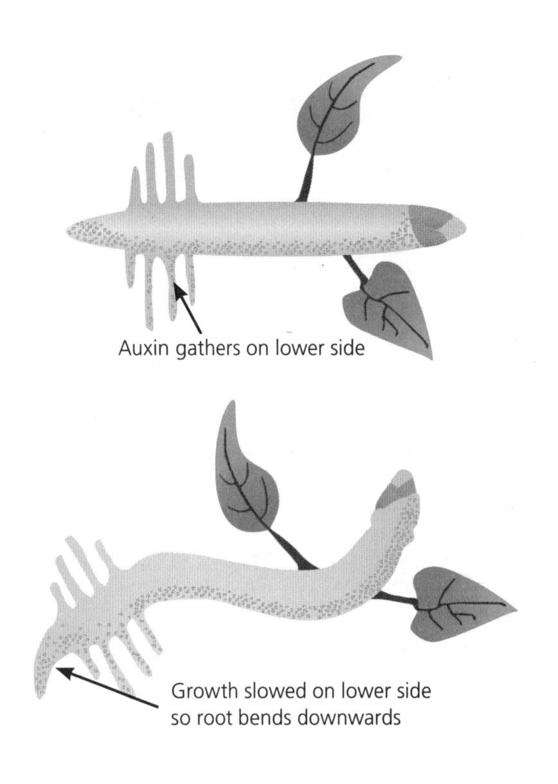

Uses of Plant Hormones

Plant hormones can be used in agriculture and horticulture. They can be used to:

- **Encourage cuttings to grow roots** – cuttings are dipped in rooting compound so they will produce roots quickly.
- **Kill weeds** – most weeds have broad leaves (unlike grass or wheat, which are narrow-leaved). Selective weedkillers are plant hormones that will disrupt the growth patterns of broad-leaved plants whilst not harming the grass or the wheat.

You need to be able to evaluate the benefits of, and problems that may arise from, the use of hormones to control fertility, including IVF.

Science Today

One in seven couples in the UK experience delays in conceiving. The most common reasons include hormonal problems, blocked Fallopian tubes in women and low sperm counts in men.

There are a number of treatments available that range from hormone treatments such as FSH, which boosts egg production, to assisted conception techniques such as *in vitro* fertilisation (IVF).

IVF involves removing eggs from the woman's ovaries and mixing them with her partner's sperm or donated sperm in a laboratory. A number of eggs are then placed back in the womb.

The success rate is just 28% and many cycles of treatment are often necessary. The NHS limits couples to one cycle of treatment, and women must be between 23 and 39 and have a specific fertility problem (e.g. blocked Fallopian tubes) or must have failed to conceive for three years despite regular intercourse.

A cycle of IVF at a private clinic costs around £4000–£5000.

Benefits of IVF
• Can help a woman become pregnant • Uses woman's own eggs and partner's own sperm • Provides an alternative to adoption
Problems of IVF
• Only 28% success rate • Age restriction • Uses NHS resources • Costly • Increases expectation for babies on demand

You need to be able to evaluate the use of plant hormones in horticulture.

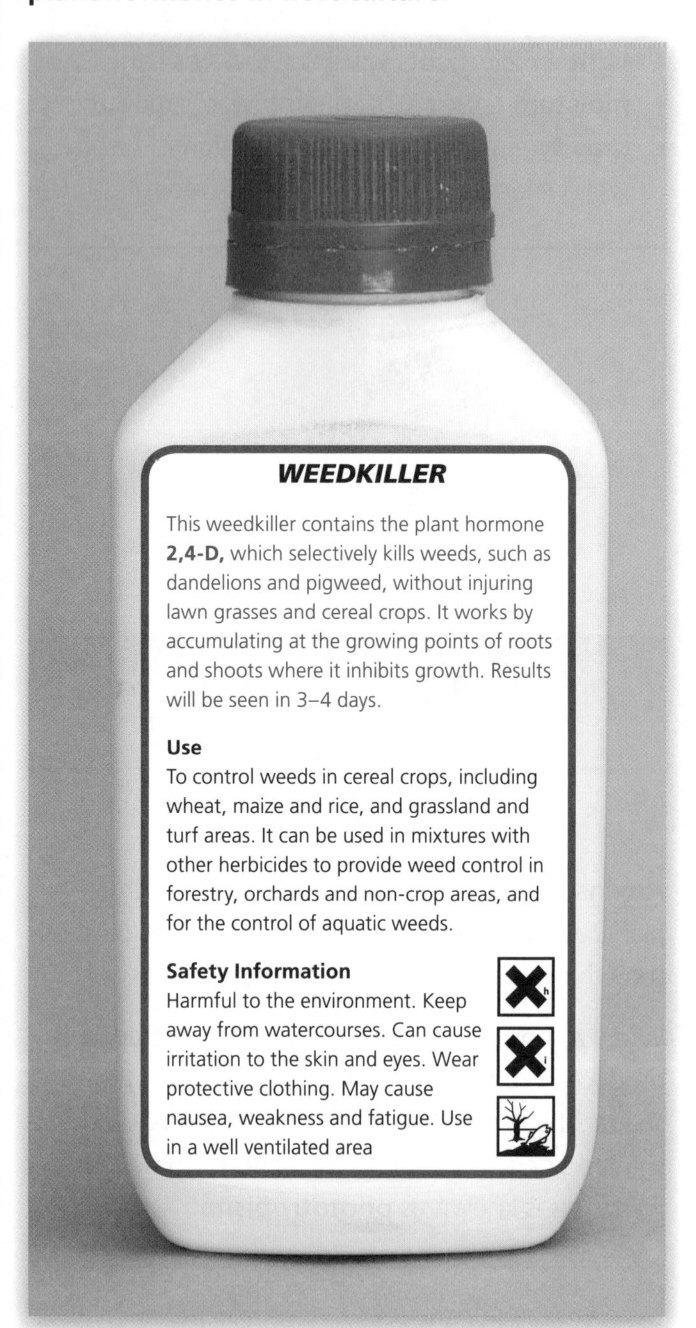

Benefits
• Kills weeds • Fast acting • Many applications
Problems
• Harmful to wildlife • Harmful to humans

B1.3 The use and abuse of drugs

Drugs are used in medicine to cure illnesses and diseases. Some drugs are used recreationally by people who like the effects they create. To understand this, you need to know:

- how medical drugs are developed and tested
- why drugs are used recreationally
- why some athletes take performance-enhancing drugs
- that people cannot make sensible decisions about drugs unless they know their full effects.

Developing New Drugs

Drugs are chemical substances that alter the way the body works. Scientists are continually developing new drugs, which need to be thoroughly tested and trialled for toxicity, efficacy and dose before use.

The flow chart shows the stages in developing a new drug.

New drug made in a laboratory

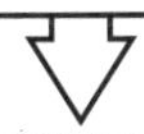

Tested in laboratory using cells, tissue culture and live animals for toxicity

Drug tested on healthy volunteers in clinical trials. Often very low doses of drug are given in initial trials to test for toxicity

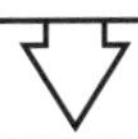

If drug found to be safe, there are further clinical trials on humans to find the optimum dose for the drug. In double blind trials some patients are given a **placebo**, which does not contain the drug. Neither patients nor doctors know who is given the placebo until the trial finishes.

Thalidomide

Thalidomide is a drug that was developed as a sleeping pill. It was tested and **approved** for this use.

It was then found to be effective in relieving morning sickness in pregnant women, but it had **not been tested** for this use. Many babies born to mothers who took the drug had severe limb abnormalities, so it was **banned**.

As a result, drug testing became much more rigorous. More recently, thalidomide has been used successfully in the treatment of **leprosy** and other diseases.

Recreational Use of Drugs

Some people use drugs recreationally because they like the effects the drug has on them. Some recreational drugs are **legal**, e.g. nicotine and alcohol. Some recreational drugs are **illegal**, e.g. cannabis, heroin and ecstasy.

Misuse of these drugs can adversely affect the heart and circulatory system. Some recreational drugs are more harmful than others.

Legal Drugs

Illegal Drugs

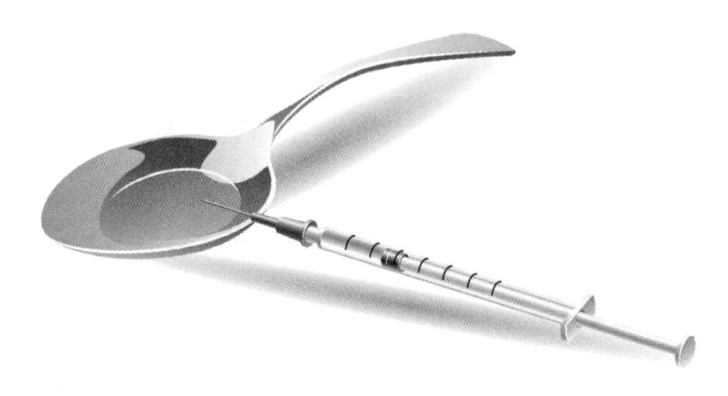

The overall impact of legal drugs on health is much greater than the impact of illegal drugs because far more people use legal drugs. For example:

- the link between smoking cigarettes and developing respiratory illnesses and lung cancer is well established
- excessive alcohol consumption and binge drinking causes damage to the liver, which can eventually result in death.

Drugs alter chemical processes in people. Therefore, people may become dependent or addicted to them and may suffer withdrawal symptoms if they do not have them. Symptoms may be psychological or physical, e.g. paranoia, sweating, vomiting. Heroin and cocaine are very addictive drugs.

Cannabis

Cannabis is an illegal drug used for **recreational** purposes by many people. It is a relaxant and can cause hallucinations. Other potentially harmful side effects include bronchitis, emphysema, lung cancer and mental illness.

Smoking cannabis has proved to be a beneficial painkiller to people suffering from a variety of medical conditions including HIV, multiple sclerosis, arthritis and cancer.

Performance Enhancers

Performance-enhancing drugs are any substances used to increase a particular skill set. For example, steroids are performance-enhancing drugs that are taken to build mass and strength of muscles and / or bones.

Stimulants will boost body functions such as heart rate.

The use of enhancement substances in sporting events is banned. Athletes can be screened for these drugs and if found to have taken them can be banned from further participation.

You need to be able to evaluate the effect of statins on cardiovascular disease.

Example

Why does cholesterol need to be reduced?
High levels of cholesterol, especially the 'bad' type (LDLs), can cause the arteries to clog up with a thick fatty substance. The result is the narrowing of the arteries that take blood to heart muscle (coronary arteries). This narrowing and hardening of the arteries is called atherosclerosis, and can cause heart attacks. A sufferer will feel a heavy, tight chest pain and perhaps experience breathing difficulties. In very serious cases, a coronary artery may become blocked by a blood clot (thrombosis). This can cause severe pain and is life-threatening.

What are statins?
Statins are drugs that work in the liver to reduce the manufacture of cholesterol. They have been found to be the most effective drugs for lowering LDL levels.

How long do you take statins for?
Statins are taken for life. They can reduce the chance of having a heart attack or a stroke by up to a third, and can increase the life expectancy of a person with a history of high cholesterol when taken long term.

Are there any side effects?
Statins may cause headaches, sickness, diarrhoea, insomnia, liver problems, stomach upsets, hepatitis and muscle aches. They cannot be taken by children, pregnant women or heavy drinkers.

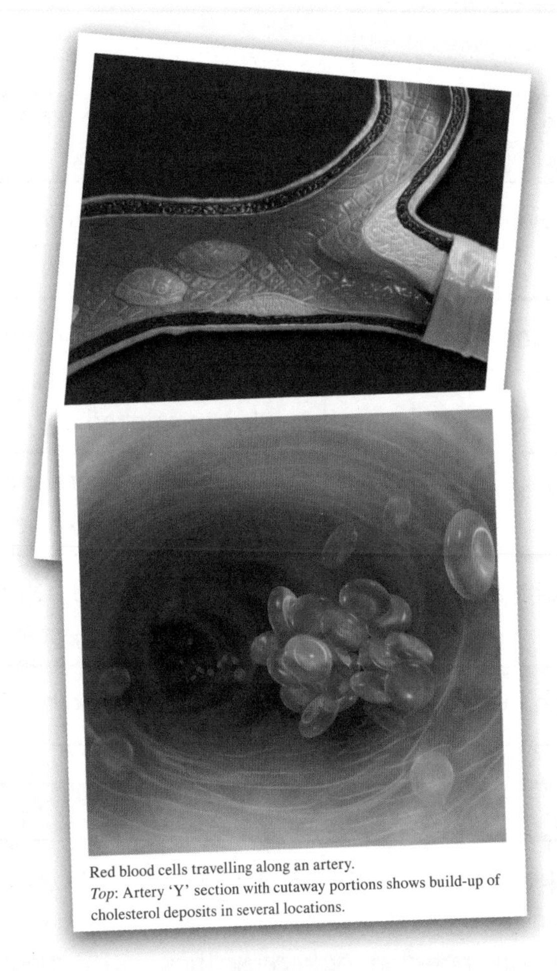

Red blood cells travelling along an artery.
Top: Artery 'Y' section with cutaway portions shows build-up of cholesterol deposits in several locations.

Is there an alternative?
If your doctor prescribes statins, you should take them. However, you can reduce the need for them if you stop smoking, eat a healthy diet with plenty of fruit and vegetables and low fat and salt content, and exercise regularly: moderate exercise three times a week reduces the risk of a heart attack by a third.

Benefits	Problems
• They reduce the manufacture of cholesterol. • They reduce the risk of heart attack or stroke caused by cardiovascular disease by up to a third. • They are the most effective drugs for reducing LDLs.	• Statins can have many side effects. • They cannot be taken by children, pregnant women or heavy drinkers. • Availability of statins reduces the need for people to change their unhealthy lifestyles. • They must be taken for life.

How Science Works

You need to be able to evaluate the different types of drugs and why some people use illegal drugs for recreation.

Example

The table below shows the different types of illegal drugs commonly used in the UK. Drugs are classified into classes A, B or C; class A is the most dangerous.

Drug	Class	Details of Drug	Number of People Using it Regularly
Heroin	A	Sedative. Causes severe cravings, very addictive. Smoked or injected.	43 000
Cocaine	A	Stimulant. Increases confidence, raises heart rate and blood pressure. Causes cravings, very addictive. Injected, inhaled or smoked.	755 000
Ecstasy	A	Stimulant. Gives adrenaline rush, feeling of well-being, high body temperature, anxiety. Ingested in tablet form.	614 000
Amphetamines	A / B	Stimulant. Increases heart rate. Can be very addictive and can cause paranoia. Ingested in tablet form.	483 000
Cannabis	B	Relaxant. Can cause hallucinations. Smoked.	3 364 000

Types of Drug Use	Problems
Experimental: when people try a drug for the first few times out of curiosity, boredom or because their friends are doing it. Some do it to rebel against authority.	• A person is just as likely to have a bad reaction from trying a drug the first time as a person who uses drugs for recreation regularly. • It could lead to recreational use. • It could lead to trying harder drugs to get a new 'high'.
Recreational: when people use drugs in a regular but fairly controlled way. It is seen by some people as a way of relaxing.	• It could lead to dependent drug use. • It could lead to using harder drugs to get a new 'high'. • A person risks a bad reaction to drugs every time they use them. • It could lead to mental problems.
Dependent: when people are addicted to drugs and need the feeling it gives them. It is also physically addictive – the body needs the drug.	• People cannot function without the drug. • People suffer withdrawal symptoms if they do not have the drug. • A person risks a bad reaction to drugs every time they use them. • It could lead to mental problems.

When using illegal drugs you can never be certain of what you are taking. Substances such as glucose, milk powder, brick dust and flour may be added to heroin and cocaine to add bulk.

Even if the drug is pure and hasn't been adulterated, a harmful substance is still being put into the body so there is always a risk of death.

You need to be able to evaluate claims about the effect of non-prescribed drugs on health and consider possible progression from recreational drugs to harder drugs.

Example

New concerns over cannabis use

A leading report published yesterday has sparked a fresh wave of concern about the dangers of using cannabis.

Concern about cannabis use is not isolated to the 21st century.

In Egypt in the 8th century, laws were introduced which prohibited the use of 'hemp drugs' (cannabis). In the 19th century, a large scale investigation into the health effects (physical and mental) of cannabis use was set up by the Indian Hemp Drugs Commission, who concluded that the link between cannabis and mental injury was complex. In recent years, research in Scandinavia has linked cannabis to severe mental illness.

The new report backs up these claims that cannabis is bad for your health, and suggests that in many cases users go on to use harder drugs, which can be extremely addictive.

1

The report claims that cannabis contains around 400 chemicals that are known to affect the brain. Cannabis smoke also contains tar and many of the same cancer-causing chemicals as tobacco.

It claims that cannabis use affects blood pressure, (causing fainting), or more severe problems for people with heart and circulation conditions. Although cannabis is not as addictive as alcohol, tobacco or amphetamines, many users do become psychologically addicted, which may then lead to them trying harder drugs.

Other reports, however, have said that cannabis, unlike harder drugs, does not lead to major health problems, and few deny that it can have beneficial effects for patients suffering from conditions such as HIV, multiple sclerosis and cancer.

The recreational use of cannabis is still illegal in Britain, although up to half of all young people have tried it.

2

Is Smoking Cannabis Bad for your Health?

Yes

- Cannabis has a higher concentration of cancer-causing substances (**carcinogens**) than tobacco.
- It has a higher tar content than tobacco so can lead to bronchitis, emphysema and lung cancer.
- It can increase the risk of fainting since it disrupts the control of blood pressure.
- It can lead to mental illness.
- It contains more than 400 chemicals. The main one that affects the brain is known as THC.
- People with heart and circulation disorders or mental illness can be adversely affected by it.
- It may be psychologically addictive.

No

- Its effects are beneficial to patients suffering from various medical conditions including HIV, multiple sclerosis and cancer.
- Unlike harder drugs, a government report has suggested that high use of cannabis is not associated with major health or sociological problems.
- Cannabis is less addictive than amphetamines, tobacco or alcohol, and does less harm to the body.

Can Smoking Cannabis Lead Users to Harder Drugs?

Yes

- Cannabis may be a 'gateway' drug to more addictive and harmful substances such as heroin and cocaine.

No

- Many cannabis smokers never use any harder drugs.

You need to be able to evaluate the use of drugs to enhance performance in sport and to consider the ethical implications of their use.

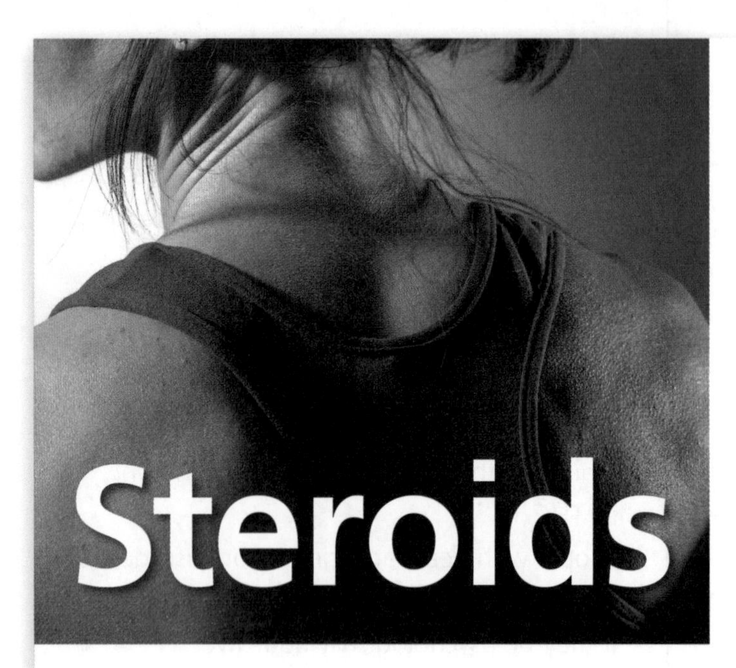

Steroids are drugs that are usually synthesised from the male reproduction hormone, testosterone. When used appropriately for medical purposes, steroids are safe and effective and can be prescribed to mitigate cancer and AIDS symptoms or to help patients gain weight after severe illness, injury or continuing infection. Steroids promote the development of muscle tissue in the body, with an associated increase in strength and power, so they have been used by athletes to improve performance in sport. However, they are controlled substances in many countries and have been banned by many sporting bodies, including the International Olympic Committee, because of their danger to health and the potential for competitive advantage. Possible harmful effects include damage to the liver and heart, cancer and male impotence. However, not everyone believes their use should be banned in sport. Here is what some people say.

Mother of teenage athlete
It's not a matter of whether it is or isn't cheating. Young people look to the professionals as role models. There is a danger that younger athletes will follow the example of the professionals without being aware of the possible dangers.

Fitness trainer
Laser surgery and specially designed contact lenses have been used by golfers to improve their vision. Cyclists have specially engineered machines and technology has even developed body hugging shirts for rugby players to help them evade tackles. Were the sporting triumphs of these competitors devalued because of the help technology had afforded them? Are any of them cheating less than an athlete taking drugs to help them recover faster from a strain or injury?

Doctor
Performance-enhancing drugs are safe if used carefully. Unfortunately, the one thing that prevents athletes from getting proper medical supervision is the fact that they are illegal.

Athlete
It's one thing to say that athletes have a choice whether they take drugs or not – but if performance enhancers were legal, then there would be huge pressure on athletes to take them to be in with a chance of winning.

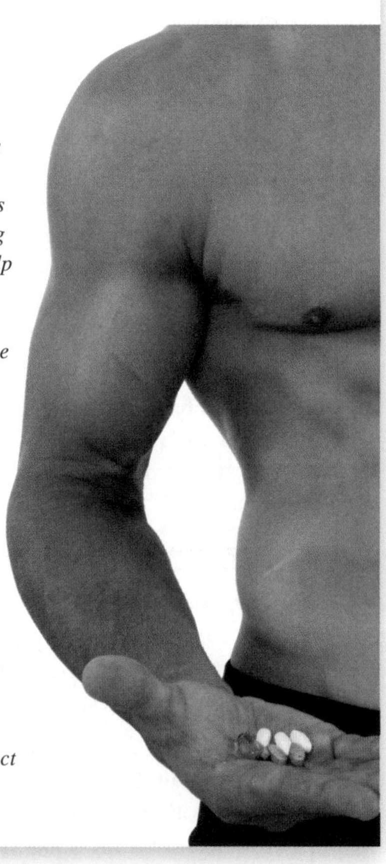

Summary:
- Steroids have medical uses.
- Steroids as performance enhancers in sport are banned.
- Side effects include damage to the heart and liver.
- Some people think steroids should be allowed in sport because it is no different to using other technological developments to improve performance.
- Some people think that if steroids were allowed there would be less risk because their use could be medically monitored.
- Young athletes may end up taking steroids because they see professional athletes taking them.
- Young athletes may end up taking steroids because they are pressurised to continually improve.

B1.4 Interdependence and adaptation

Organisms are well adapted to survive in their normal environments. To understand this, you need to know that:

- population size depends on a variety of factors
- changes in the environment may affect the distribution and behaviour of organisms.

To survive and reproduce, organisms need a supply of materials from their surroundings and from the other organisms there.

Competition

Organisms compete with each other for space / light, food and water.

Factor	Plants	Animals
Space / light	Need room to spread leaves to obtain light for photosynthesis	Need space to breed and compete for a mate. Also territory to hunt in
Food	Absorb nutrients from the soil	Herbivores compete for vegetation, and carnivores compete for their prey
Water	Absorb water by their roots	Need water in order to survive

A **population** is the total number of individuals of the same species that live in a certain area, e.g. the number of field mice in a meadow.

When organisms compete in an area or habitat, those that are better adapted to the environment are more successful and usually exist in larger numbers. This often results in the complete exclusion of other competing organisms.

Adaptations

Adaptations are special features or behaviour that make an organism particularly well-suited to its environment. Adaptating is part of the evolutionary process that 'shapes life', so a habitat is populated by organisms that excel there.

Adaptations increase an organism's chance of survival; they are biological solutions to an environmental challenge.

For example:

- Some plants have thorns to prevent animals from eating them (e.g. roses and cacti).
- Some organisms have developed poisons and warning colours to deter **predators** (e.g. blue dart frogs).
- Some organisms are adapted to live in areas of high salt concentration, whilst some organisms can survive extremes of temperature or pressure. Such organisms are termed **extremophiles**, e.g. the Pompeii worm, found in the Pacific Ocean, can survive temperatures of up to 80°C in deep sea vents.

Microorganisms have a wide range of adaptations, enabling them to live in a wider range of conditions.

Blue Dart Frog

Animal Adaptations

Animals and plants may be adapted for survival in the conditions where they normally live, e.g. deserts or the Arctic. They may be adapted by different methods, for example:

- changes to **surface area**
- thickness of insulating coat
- amount of body fat
- camouflage.

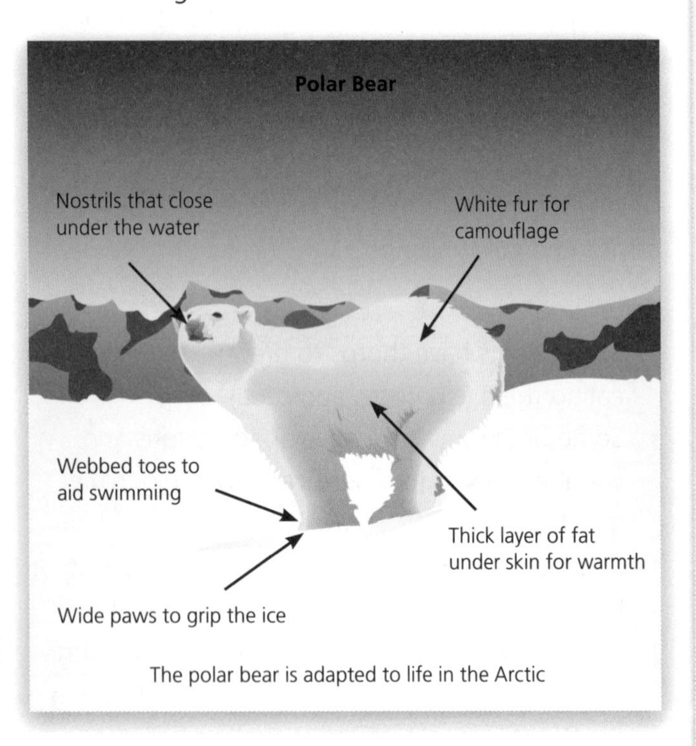

The polar bear is adapted to life in the Arctic

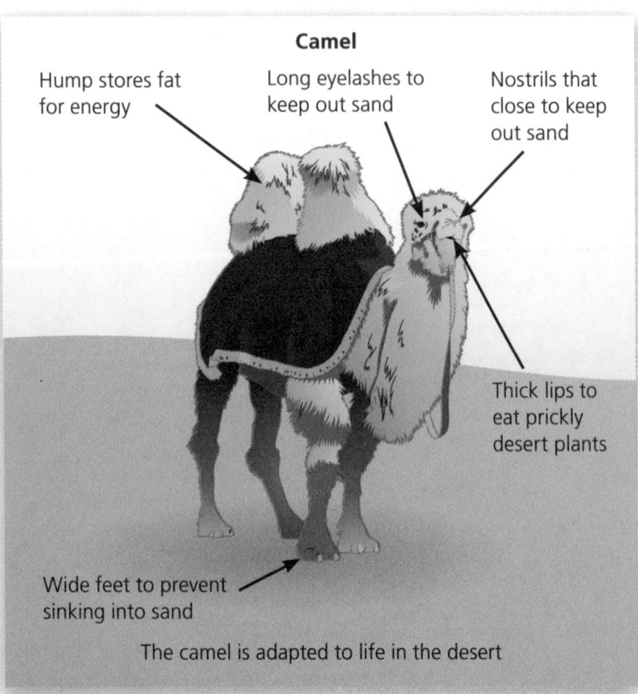

The camel is adapted to life in the desert

Plant Adaptations

Plants lose water vapour through their leaves and may have adaptations to enable them to survive in dry environments.

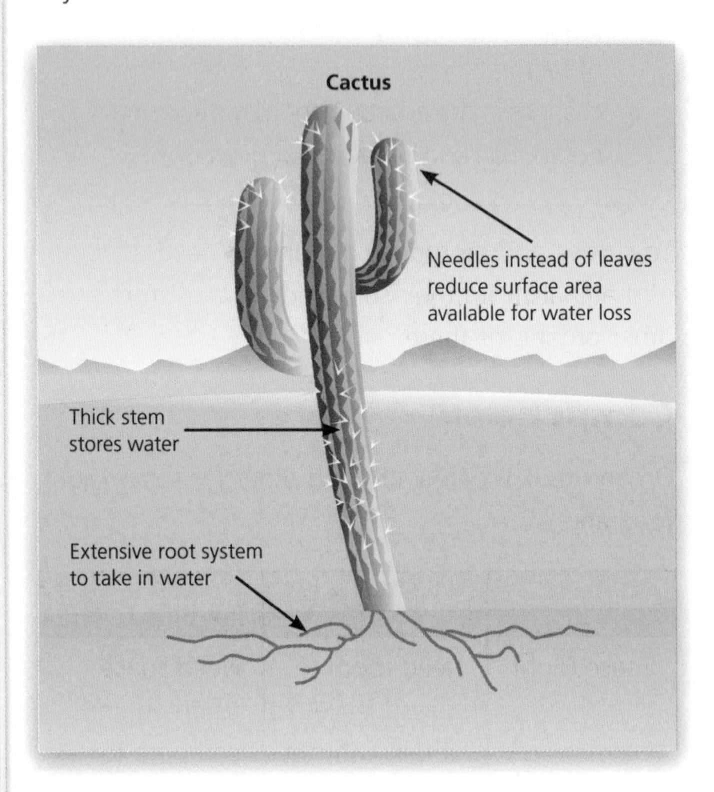

Environmental Change

Animals and plants are subjected to environmental changes that may be caused by living or non-living factors.

Non-living changes might include changes to average temperatures or the amount of rainfall. This will affect the distribution of living organisms. For example:

- crops that previously grew well in a dry area may not thrive in a wetter environment
- the numbers of pollinating insects (e.g. bees) may decrease in cooler weather, which may in turn affect the distribution of future crops.

Living changes that affect the distribution of organisms include the introduction of new predators, new diseases and migration.

Plant populations are also affected by grazing (by herbivores) and disease.

Indicator Organisms

Living organisms can be used as indicators of pollution.

Lichens

Lichens are organisms that are affected by sulfur dioxide pollution in the air. Different types of lichens have different levels of sensitivity, therefore, the types and numbers of lichens in an area can give a good indication of sulfur dioxide levels in the air.

In towns and industrialised areas there is often a high level of sulfur dioxide. There are likely to be low numbers of lichens and only a few varieties found growing on trees and walls in these areas.

In areas where there are lower levels of sulfur dioxide (e.g. rural areas) there is likely to be a larger number and wider variety of lichens found.

Few lichens indicate high concentration of sulfur dioxide in the air

Many lichens indicate low concentration of sulfur dioxide in the air

Invertebrate Animals

Invertebrate animals can also be used as indicator organisms. When rivers or lakes become polluted, the level of dissolved oxygen in the water falls. Many organisms will die when oxygen levels fall, however some invertebrate organisms can tolerate polluted water where oxygen levels are low. There are also some organisms that are only found in 'clean' water where dissolved oxygen levels are high.

The numbers and types of organisms found in a lake or river can be used to indicate the level of oxygen in the water.

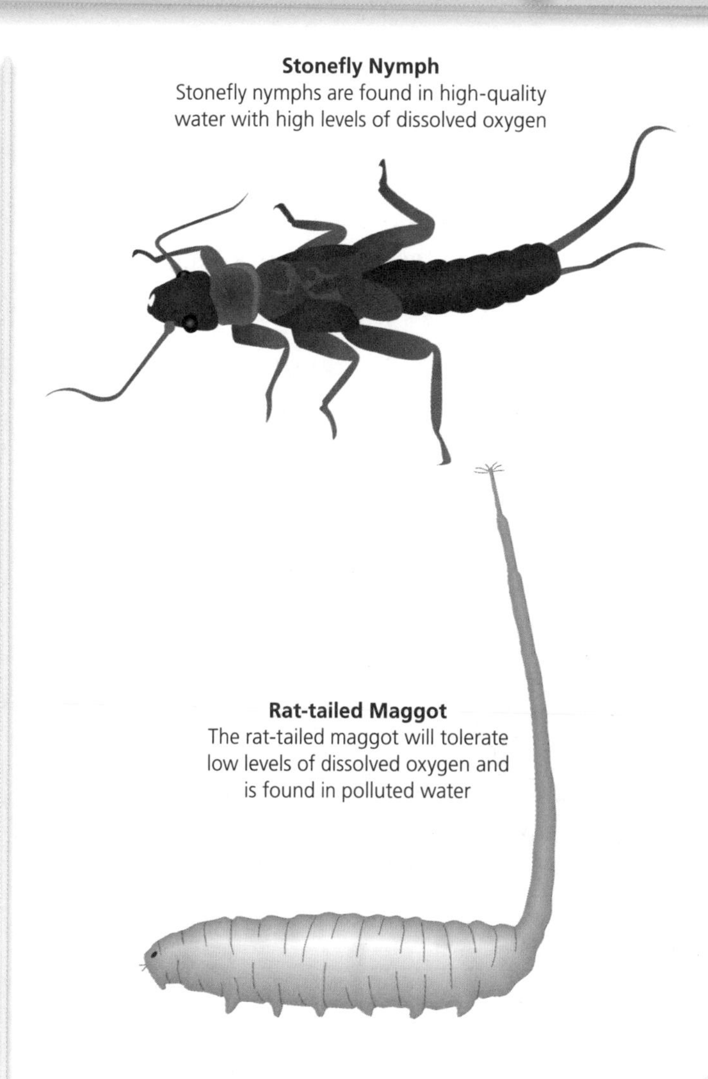

Stonefly Nymph
Stonefly nymphs are found in high-quality water with high levels of dissolved oxygen

Rat-tailed Maggot
The rat-tailed maggot will tolerate low levels of dissolved oxygen and is found in polluted water

Measuring Environmental Change

Environmental change can also be measured using non-living indicators – for example, equipment to measure temperature, rainfall and oxygen levels in water.

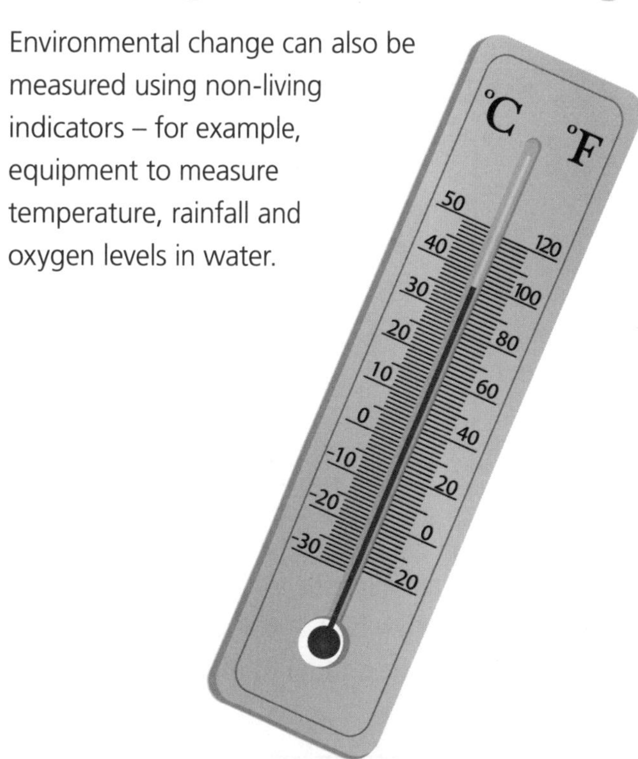

You need to be able to suggest how organisms are adapted to the conditions in which they live.

Example

A group of students from a school in North Yorkshire set off to the coast to investigate the effect of waves on organisms on the seashore.

They decided to base their study on the width and height of limpets in two rocky bays – Runswick Bay and Robin Hood's Bay – to see how they were adapted to their environment.

The rocks at Runswick Bay were large boulders. They were exposed to the force of the waves on one side but sheltered on the other side.

At Robin Hood's Bay, the sandstone outcrops were shelved as they stretched gradually out to sea.

RUNSWICK BAY

	Width of base of limpet shell (cm)	Height of limpet shell (cm)
Limpet 1	1.4	1.4
Limpet 2	2.4	1.5
Limpet 3	2.5	1.6
Limpet 4	2.4	2.5
Limpet 5	2.6	2.6
Limpet 6	3.5	2.1
Limpet 7	1.3	2.1
Limpet 8	2.3	1.9
Mean	2.3	2.0

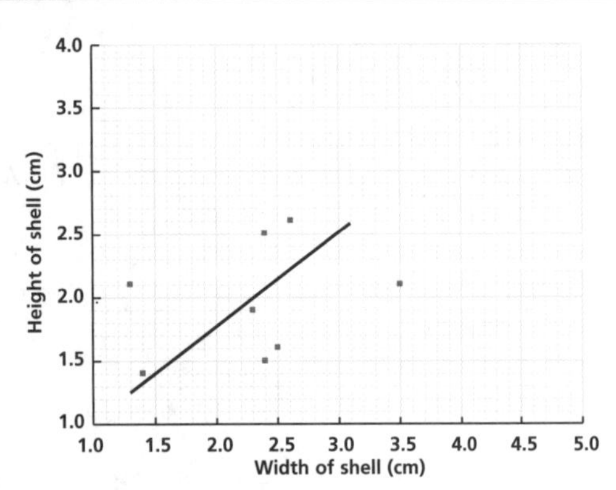

The data showed that at Runswick Bay, where the boulders are exposed to the waves, the limpet shells had a broad base compared to their height. This allowed them to resist the action of the waves.

ROBIN HOOD'S BAY

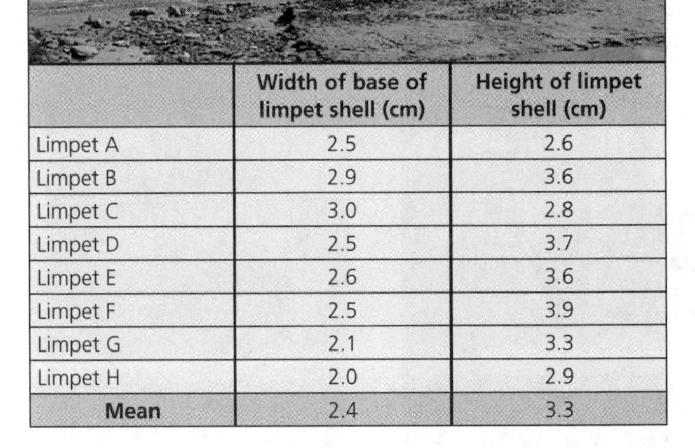

	Width of base of limpet shell (cm)	Height of limpet shell (cm)
Limpet A	2.5	2.6
Limpet B	2.9	3.6
Limpet C	3.0	2.8
Limpet D	2.5	3.7
Limpet E	2.6	3.6
Limpet F	2.5	3.9
Limpet G	2.1	3.3
Limpet H	2.0	2.9
Mean	2.4	3.3

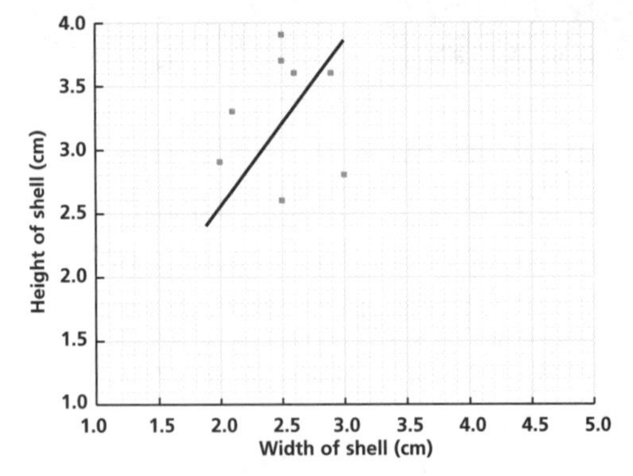

The data from Robin Hood's Bay showed that the limpet shell bases were quite narrow compared to the height. The wave action here is not as forceful as at Runswick Bay so the limpet does not require such a broad base to hold it on the rocks.

You need to observe the adaptations of a range of organisms from different habitats and develop an understanding of the ways in which adaptations enable organisms to survive.

The Hagfish

Hagfish live in the sea. When threatened, they produce large amounts of sticky slime. The slime deters predators. They clean themselves afterwards by tying themselves in a knot, scraping off the slime as they form the knot.

Black Vultures

The black vulture is a scavenging bird, feeding on the flesh of dead animals. The vultures' bare head is an interesting adaptation that allows them to stick their head within the carcass of a dead animal without fear of soiling feathers. The beak is long and hooked, which is an adaptation for tearing flesh. Black vultures are efficient gliders, an adaptation needed for scanning large territories in search of a food source.

The Aye Aye

The aye-ayes' most unique adaptation is their long, thin fingers that are used to find their food sources.

Their third finger can grow up to six inches long and is used for tapping on wood to find insects and then extracting them from the wood when found. Their oversized ears also help because when they tap they listen for vibrations to see if the wood contains an insect or food source.

Aye-ayes have very large front teeth, which they use for gnawing at wood and for breaking nuts and fruits when they eat them. Aye-ayes rely on climbing and have claws that help them cling to trees.

Star-nosed Mole

The star-nosed mole lives in wet areas and eats small invertebrates, aquatic insects and worms. It is a good swimmer and can forage along the bottom of streams and ponds. It digs shallow surface tunnels that often exit underwater. This mole, like all moles, has very poor sight.

Black Vulture

Echidna

The star-nosed mole has thick, black, water-repellent fur and a thick tail that acts as a storage area. It has 22 pink fleshy tentacles on the end of its nose, which identify food such as worms and insects by touch. It also has powerful claws for digging.

Echidna

The echidna is shy and non-confrontational – its spines are actually fur and it protects itself by curling into a ball to protect its soft belly. It has a long, moist snout and an even longer tongue, which it uses to feast on termites. Echidnas use their strong claws to dig into ant nests and termite mounds. Echidnas have no teeth, instead they crush insects between their tongue and the roof of the mouth. Echidnas have a stomach with a special lining that can accommodate a lot of dirt.

You need to be able to evaluate data concerned with the effect of environmental changes on the distribution and behaviour of living organisms.

Example

A group of students wanted to investigate the distribution of purple vetch plants near a new motorway. They made a 20-metre **line transect** as shown in the diagram. They placed a quadrat every two metres along the transect and counted the number of purple vetch plants inside the quadrat. Their results are shown on the scatter diagram.

The graph shows that as the distance from the motorway increased, the number of purple vetch plants increased. The students concluded that exhaust fumes from traffic on the new motorway were causing high levels of pollutants that were preventing the vetch from growing next to the motorway.

In this investigation the evidence gathered is limited by the method used because the students did not measure the levels of pollutants along the transect; they only carried out one line transect at one point on the motorway and on one side of the motorway. There were a number of factors not controlled, any one of which could have affected the results.

For example:

- aspect of land (flat/hilly)
- light or shade
- moisture content of soil
- mineral content of soil
- competition from other plants
- grass cutting/trampling.

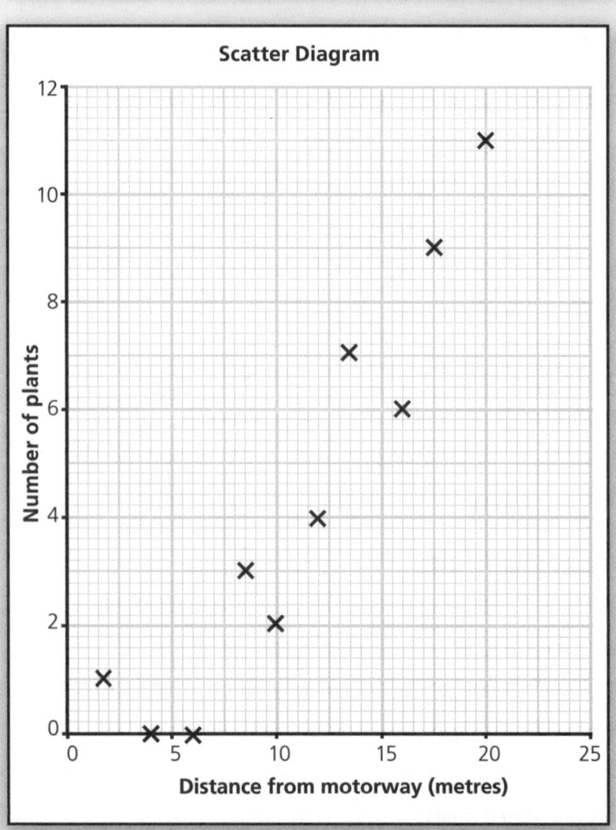

Energy and Biomass in Food Chains

B1.5 Energy and biomass in food chains

There are changes to energy and biomass at different stages in the food chain. To understand this, you need to know:

- what pyramids of biomass show
- how the amount of energy available at each stage of the food chain is reduced.

Food Chains

Radiation from the Sun is the source of energy for all communities of living organisms.

In green plants and algae, photosynthesis captures a small fraction of the solar energy that reaches them. This energy is stored in the substances that make up the cells of the plant. This energy can be passed on to organisms that eat the plant. This transfer of energy can be represented by a **food chain**.

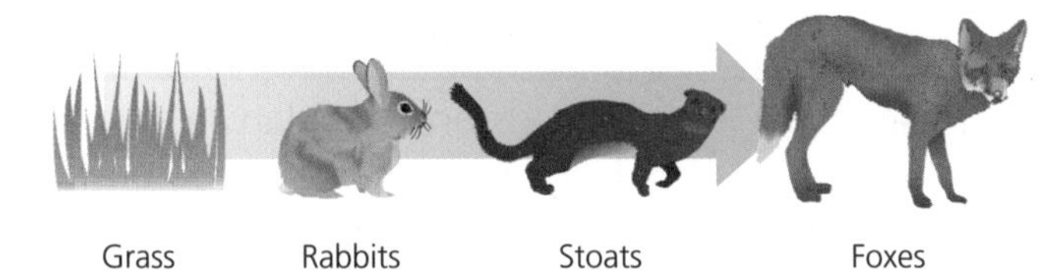

Transfer of Energy and Biomass

Biomass and energy are lost at every stage of a food chain because:

- materials and energy are lost in an organism's faeces (waste).
- energy released through respiration is used up in movement and lost as heat energy.

This is particularly true in warm-blooded animals (birds and mammals), whose bodies must be kept at a constant temperature, higher than their surroundings.

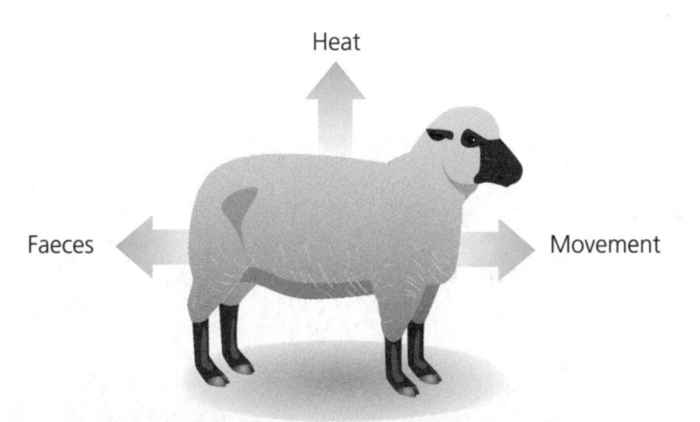

Pyramids of Biomass

The mass of living material (**biomass**) at each stage of a food chain is less than it was at the previous stage. The biomass at each stage can be drawn to scale and shown as a pyramid of biomass.

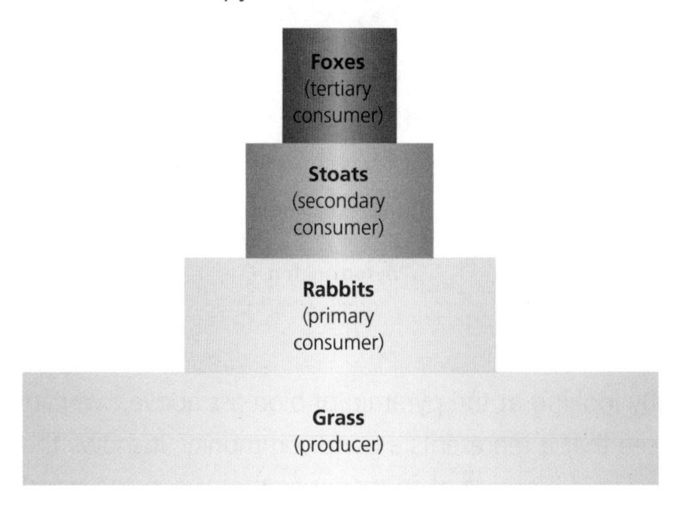

The fox gets the last tiny bit of energy and biomass that is left.

- Only a fraction of the Sun's energy is captured by the producers. Much of the biomass remains in the root system and so does not get eaten.
- Rabbits run, mate, excrete, generate heat and pass on only a tenth of the energy they get from grass. A lot of biomass is lost in droppings (faeces).
- Stoats run, mate, excrete, generate heat and pass on only a tenth of the energy they get from the rabbits. A lot of biomass is lost as faeces.
- Foxes, as tertiary consumers, only receive a small amount of the energy captured by the grass at the bottom of the pyramid.

You need to be able to interpret pyramids of biomass and construct them from appropriate information.

Example

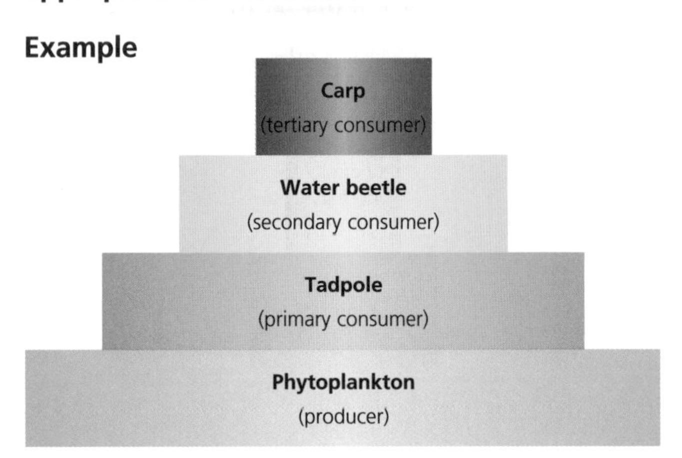

By looking at the pyramid of biomass above, we can see that it represents a pond community. It shows the relative biomass of four species that live in the pond and it also shows us what eats what in the food chain.

We can see that the **producer** is phytoplankton (a plant). This provides food and energy for the tadpoles, which are eaten by water beetles, which in turn are consumed by carp.

We can also see that the biomass of living material reduces as you go further up the pyramid (more levels of **consumers**). This is because biomass is lost as waste products, e.g. faeces, carbon dioxide and water.

The table provides details of other organisms in the pond.

Species	Producer / Feeds On
Phytoplankton	Producer
Elodea (water plant)	Producer
Algae	Producer
Water beetle	Feeds on water plants, amphibian larvae and small aquatic insects
Carp	Feeds on insect larvae and snails
Tadpole	Feeds on water plants
Pond snail	Feeds on water plants and algae
Mosquito larvae	Feeds on algae and protozoa
Leech	Feeds on pond snails
Stickleback	Feeds on insect larvae and leeches

To construct a pyramid of biomass from this information there are a few basic rules to follow:

- Biomass is the mass of living material.
- The producers are placed at the bottom.
- The pyramid represents the total biomass at each stage, not the number of organisms.

B1.6 Waste materials from plants and animals

Trees and animals produce waste, which they need to get rid of. To understand this, you need to know:

- how microorganisms play an important part in decomposing material
- how the same material is recycled over and over again and can lead to stable communities
- how the carbon cycle works.

Recycling the Materials of Life

Living things remove materials from the environment for growth and other processes, but when these organisms die or excrete waste, these materials are returned to the environment.

The key to all this is the **microorganisms** that break down the waste and the dead bodies. This **decay** process releases substances used by plants for growth.

Microorganisms digest materials faster in warm, moist conditions where there is plenty of oxygen.

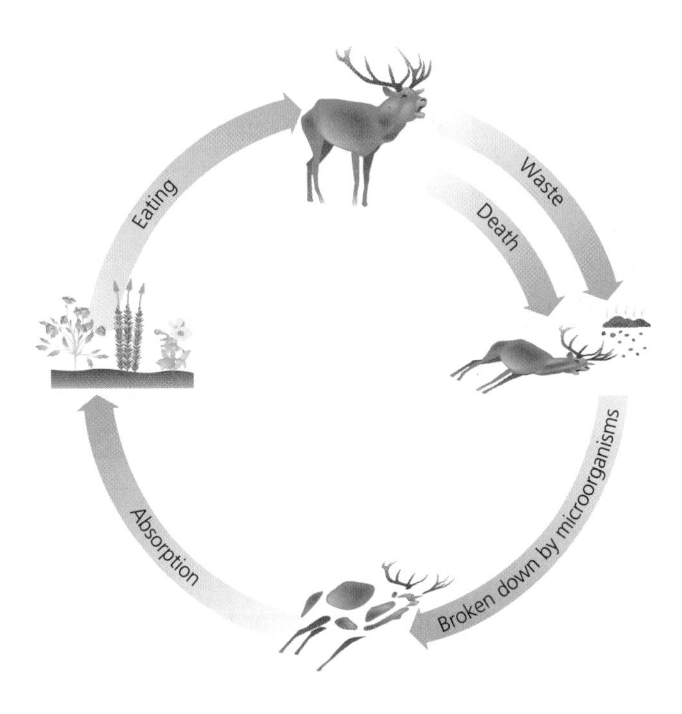

In a stable community, the processes that remove materials are balanced by processes that return materials.

The Carbon Cycle

The constant recycling of carbon is called the **carbon cycle**.

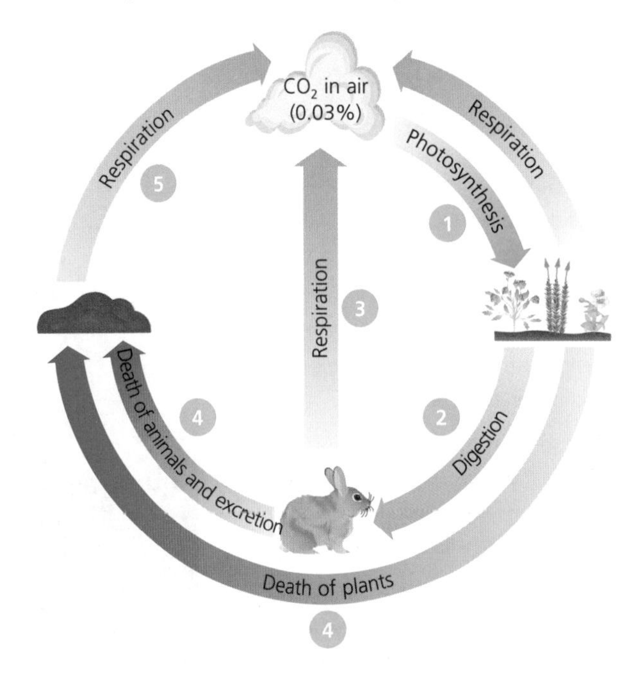

1. CO_2 is removed from the atmosphere by green plants and algae to produce glucose by photosynthesis. Some CO_2 is returned to the atmosphere by plants and algae during respiration.
2. The carbon obtained by photosynthesis is used to make carbohydrates, fats and proteins in plants. When the plants are eaten by animals this carbon becomes carbohydrates, fats and proteins in animals.
3. Animals respire, releasing CO_2 into the atmosphere.
4. When plants and animals die, other animals and microorganisms feed on their bodies, causing them to break down.
5. As the **detritus** feeders and microorganisms eat the dead plants and animals, they respire releasing CO_2 into the atmosphere.

By the time the microorganisms and detritus feeders have broken down the waste products and dead bodies of organisms and cycled the materials as plant nutrients, all the energy originally captured by green plants has been transferred. Combustion of wood and fossil fuels also releases carbon dioxide into the atmosphere.

You must be able to evaluate the necessity and effectiveness of schemes for recycling organic kitchen or garden waste.

Why composting is important

Some 38% of household rubbish is kitchen or garden waste and most of this ends up in landfill sites.

When green waste ends up in landfill sites, it breaks down giving off methane, a 'greenhouse gas' that makes a big contribution to global warming.

Composting waste at home doesn't produce methane, so not only are you making free food for your soil but you are also helping to reduce climate change.

Making your own compost has many benefits. It provides nutrients for plants, helps soil to retain moisture and reduces the demand for peat-based composts.

What can I compost at home?

YES PLEASE ✓

- Fruit and vegetable peelings
- Tea bags
- Paper towels / tissues
- Crushed egg shells
- Grass cuttings

NO THANKS ✗

- Meat
- Fish
- Dairy products
- Cat or dog excrement
- Diseased plants

Some frequently asked questions

Why can't I put cooked vegetables into my home composter?

They may generate smells and attract vermin to your garden.

What can I do with food waste that I can't compost at home?

Many councils now provide householders with food waste collection bins. These are collected every week or fortnight. You can put cooked food and meat and dairy products in these.

What does the council do with this waste?

It is anaerobically digested in a sealed system to produce biogas and organic fertiliser. The biogas can be used to provide electricity whilst the fertiliser is used on farm land to grow crops.

Do all councils offer kitchen waste recycling schemes?

About 90% of local councils offer garden waste collection and about 38% offer food waste collections.

Benefits	Problems
• Less methane produced, which helps to reduce global warming • Does not take up precious space in landfill sites • Provides plants with nutrients • Cuts down on buying peat-based composts • Most councils offer garden-waste collection • Councils can produce biogas and fertilisers from the waste	• Many food waste items cannot be composted at home • Food waste left lying around could attract vermin • Fewer than half of councils offer food-waste recycling schemes

B1.7 Genetic variation and its control

There are differences between different species of plants and animals. There are also differences within species. Scientists can manipulate genes to produce plants and animals with the characteristics they want. To understand this, you need to know:

- about genes and chromosomes
- the differences between sexual and asexual reproduction
- the methods scientists use for producing plants and animals with the characteristics they prefer.

Genetic Information

The information that results in plants and animals having similar characteristics to their parents is carried by genes that are passed on in the sex cells (gametes) from which they develop.

The nucleus of a cell contains **chromosomes**, which are made up of a substance called **DNA**. A section of a chromosome is called a **gene**. Genes carry information that control the characteristics of an organism. Different genes control the development of different characteristics. During reproduction, genes are passed from parent to offspring (i.e. they are inherited).

A Section of One Chromosome

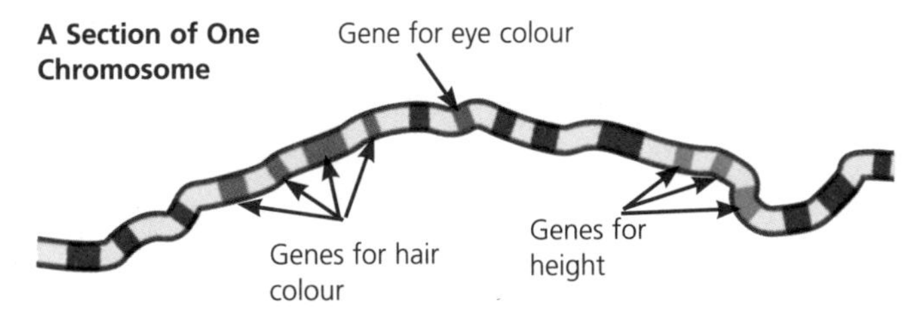

Chromosomes come in pairs, but different species have different numbers of pairs, e.g. humans have 23 pairs.

Cell with Two Pairs of Chromosomes

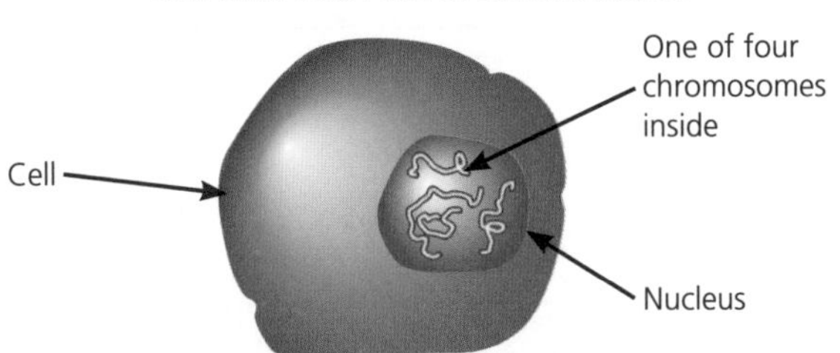

Causes of Variation

Differences between individuals of the same species are called **variation**. Variation may be due to genetic causes (i.e. the different genes that have been inherited) or environmental causes (i.e. the conditions in which the organism has developed). Variation is usually due to a combination of genetic and environmental causes.

For example, identical twins will look exactly alike at birth (genetic), but their lifestyle can alter how they look throughout their lives (environment). So if one twin has a diet high in fat and does no exercise, he will become fatter than his brother. Genetics are responsible for things such as the colour of dogs' coats.

Effect of Reproduction on Variation

During **sexual reproduction**, a sperm from a male fuses with an egg from a female. When this happens, the genes carried by the egg and the sperm are mixed together to produce a new individual.

This mixing is completely random, so produces lots of variation, even among offspring from the same parents.

Asexual reproduction means no variation at all, unless it is due to environmental causes.

Only one parent is needed and individuals who are genetically identical to the parent (clones) are produced. Bacteria reproduce asexually.

Reproducing Plants Artificially

Plants can reproduce **asexually** (i.e. without a partner) and many do so naturally. All the offspring produced asexually are **clones** (i.e. they are genetically identical).

Taking Cuttings

When a gardener has a plant with all the desired characteristics he may choose to produce lots of them by taking stem, leaf or root cuttings. Cuttings are a quick and cheap way of producing new plants. These should be grown in a damp atmosphere until roots develop.

Spider Plant Stolons

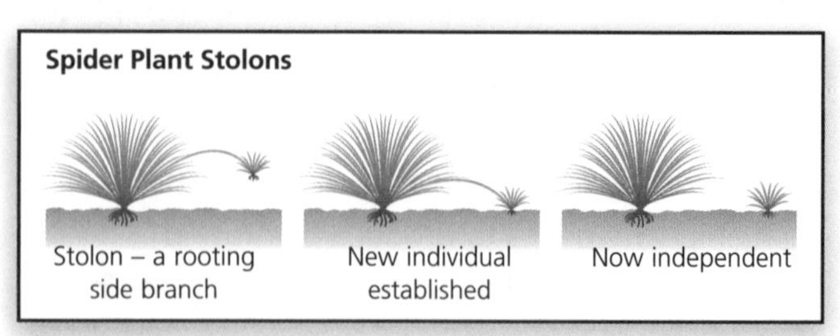

Taking cuttings

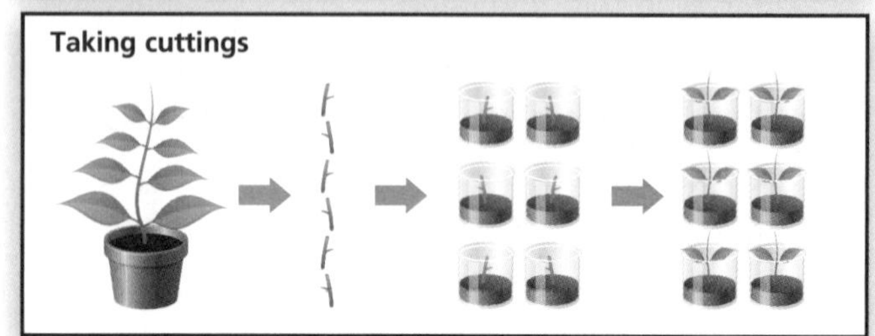

Cloning

Clones are genetically identical individuals. For example, if you have a plant that has the ideal characteristics you can clone it to produce more plants with the same desired characteristics. This is exactly what is happening in modern agriculture.

Tissue Culture

- Parent plant with the characteristics that you want is selected.
- A few cells are scraped off into several beakers containing nutrients and hormones.
- A week or two later there are lots and lots of genetically identical plantlets growing. The same can be done to these.
- This whole process must be aseptic (carried out in the absence of harmful bacteria) otherwise the new plants will rot.

N.B. The offspring are genetically identical to each other and to the parent plant.

Embryo Transplants

Instead of waiting for normal breeding cycles, farmers can obtain many more offspring by using their best animals to produce embryos that can be inserted into 'mother' animals.

- Parents with desired characteristics are mated.
- Embryo is removed before the cells become specialised…
- …then split apart into several clumps.
- These embryos are then implanted into the uteruses of sheep who will eventually give birth to clones.

N.B. The offspring are genetically identical to each other, but not to the parents.

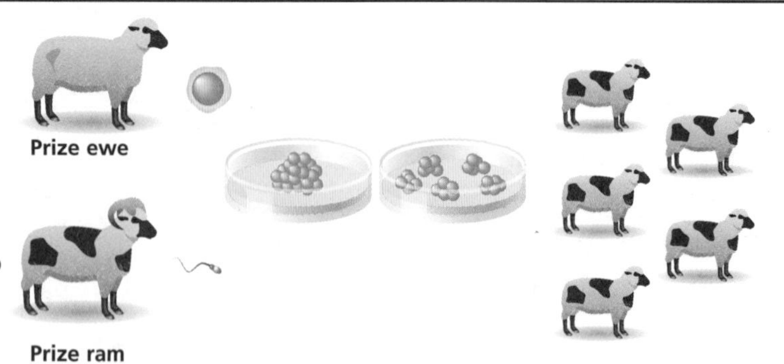

Adult Cell Cloning / Fusion Cell Cloning

- DNA from a donor animal is inserted into an empty egg cell (nucleus removed).
- An electric shock causes the egg to begin to divide to form embryo cells, which contain the same genetic information as the donor animal cell.
- When the embryo has developed into a ball of cells it is implanted into the womb of an adult female. Dolly the sheep was produced this way.

Genetic Modification of Crops

Crops that have had their genes modified by the addition of new DNA are called **genetically modified** (**GM**) crops. There are several reasons why crops may be genetically modified:

- To improve the crop yield, e.g. to produce larger tomatoes, potatoes, wheat seed-heads, more oil from oilseed rape, etc.
- To improve resistance to pests or **herbicides**, e.g. soya plants have been modified so that they are resistant to herbicides, allowing farmers to eliminate weeds without killing the crop.
- To extend the shelf-life of fast-ripening crops, e.g. tomatoes.

However, there are a number of concerns about GM crops:

- Genes from GM crops could be transferred via pollen to non-GM crops or even to other plants. For example, genes for herbicide resistance could find their way into weeds, causing 'superweeds'.
- The nectar from GM crops can harm the insects that feed on it.
- There is uncertainty about the effects of eating GM crops on human health.

Genetic Engineering

Genes can also be transferred to the cells of animals and microorganisms at an early stage to develop with the desired characteristics. These characteristics can then be passed on if the organism reproduces asexually or is cloned.

Insulin is an example of a substance produced by genetic engineering. Insulin is normally produced by the pancreas and helps to control blood glucose levels. Diabetics do not produce enough insulin and often need to inject it. The gene for human insulin can be inserted into bacteria, which then produce large quantities of genetically engineered insulin.

1. Scientists use enzymes to cut a chromosome at specific places so they can remove the precise piece of DNA they want. In this case, the gene for insulin production is cut out.

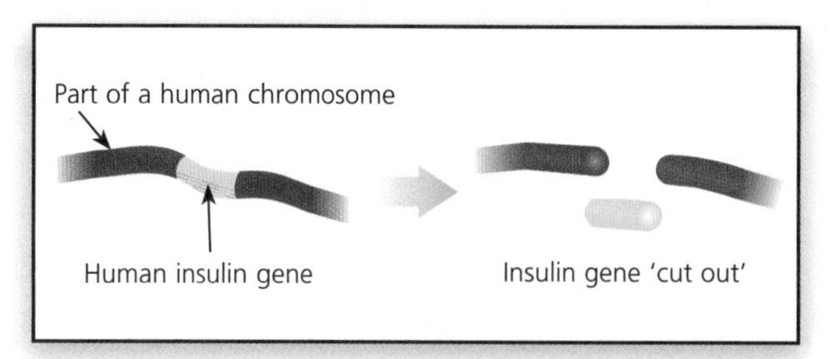

2. Another enzyme is then used to cut open a ring of bacterial DNA (a plasmid). Other enzymes are used to insert the piece of human DNA into the plasmid.

3. The plasmid is reinserted into a bacterium which starts to divide rapidly. As it divides it replicates the plasmid and soon there are millions of them, each with instructions to make insulin.

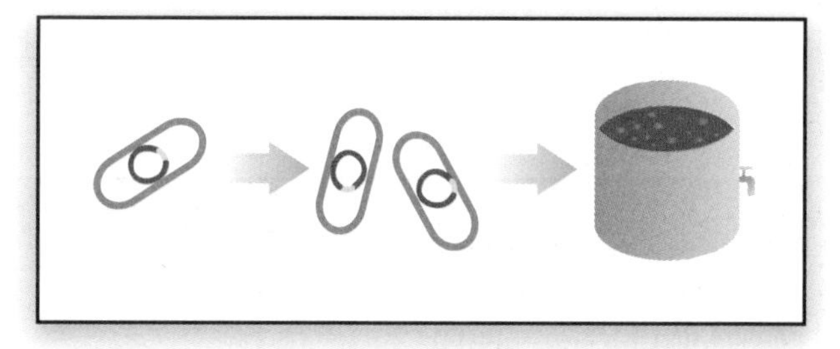

When the above process has been completed, the bacteria are cultured on a large scale and commercial quantities of insulin are then produced.

How Science Works

You need to be able to interpret information about cloning and genetic engineering techniques and make informed judgments about the economic, social and ethical issues concerning cloning and genetic engineering, including GM crops.

A basic understanding of the different types of cloning is key to taking an informed stance and making the best possible personal decisions. There are three types of cloning technologies.

Recombinant DNA Cloning	Reproductive Cloning	Therapeutic Cloning
A DNA fragment is transferred from one organism to another. This could be used to produce bacteria that will make a human protein, e.g. insulin. It can also be used to create genetically engineered crops with the desired characteristics.	Reproductive cloning is used to generate an animal that has the same nuclear DNA as another currently or previously existing animal (a twin). This can be used to produce farm animals with the desired qualities.	Therapeutic cloning, also called 'embryo cloning', is the production of human embryos for use in research. The goal of this process is not to create cloned human beings, but rather to harvest stem cells that can be used to treat disease.

Argument for Cloning Plants, Animals and Humans	Argument against Cloning Plants, Animals and Humans
• Traditional breeding methods are slow • Can predict characteristics of offspring • Allows quick response to livestock / crop shortages • Produces genetically superior stock • Elimination of diseases such as diabetes • Organ donation: a clone has matching tissue to the parent, so it would be able to donate an organ without the risk of the receiver rejecting the organ • Treatment of many degenerative diseases • Could be used to repopulate endangered species	• Loss of livelihood by traditional farmers • Risk of expression of 'unwanted' genes that adversely affect stock • May cause a backlash against cloned stock leading to market crashes • Cloning companies may have a monopoly on patent for clones • Cloning is unnatural • The fear of creating the 'perfect race' • Human clones will not have 'parents' • Cloning goes against the principles of some religions • Abnormalities may occur in a clone • Cloning does not allow natural evolution

Argument for GM Crops	Argument against GM Crops
• They are more cost-effective (manufacturers claim higher yields) • They reduce **pesticide** use (according to a US study) • They can benefit human health (they can be enriched with nutrients) • They are safe for human consumption (according to the British Medical Association). • They could help provide more food for the developing world by increasing yields • They preserve natural habitats as less land is needed for agriculture	• They increase pesticide use because farmers spray freely (according to another US study on maize) • Cross-contamination of non-GM crops could destroy the GM-free trade • They mainly benefit big GM companies • Increasing yields will not help the developing world (the problem is distribution of food, not lack of it) • Could affect wildlife since there are no weeds as a food source for animals

B1.8 Evolution

New species of organisms have developed over time. To understand this, you need to know:
- the different theories of evolution
- how new species are formed.

The Theory of Evolution

Darwin's theory of **evolution** by **natural selection** stated that all species of living things evolved from simple life forms that first developed more than three billion years ago.

This theory was only gradually accepted because:
- the theory challenged the idea that God created the Earth and everything on it
- there was insufficient evidence at the time to convince many scientists
- the mechanisms of inheritance were not known until 50 years after Darwin's theory was published.

Other theories, including that of **Lamarck**, suggested that changes occurred to an organism during its life time and these changes were then inherited. We now know that this is wrong because the genes that an organism passes to its offspring are present at birth.

Evolution is the change in a population over a large number of generations that may result in the formation of a new species, the members of which are better adapted to their environment.

There are four key points to remember:
- Individuals within a population show a wide range of variation because of the differences in their genes.
- Individuals with characteristics most suited to their environment are more likely to survive and breed successfully. This is known as 'survival of the fittest'.
- The genes that have enabled these individuals to be successful are passed on to their offspring, resulting in an improved species that has evolved by natural selection.
- Where new forms of a gene result from **mutation**, there may be a more rapid change in a species.

Studying the similarities and differences between organisms can help us to understand evolution. It can also help us to classify organisms.

Evolutionary trees are models that allow us to represent the relationships between organisms. In the tree below you can see that the first vertebrates to evolve were the fish, followed by amphibians and reptiles. Birds evolved from reptiles. Mammals were the last group of vertebrates to evolve.

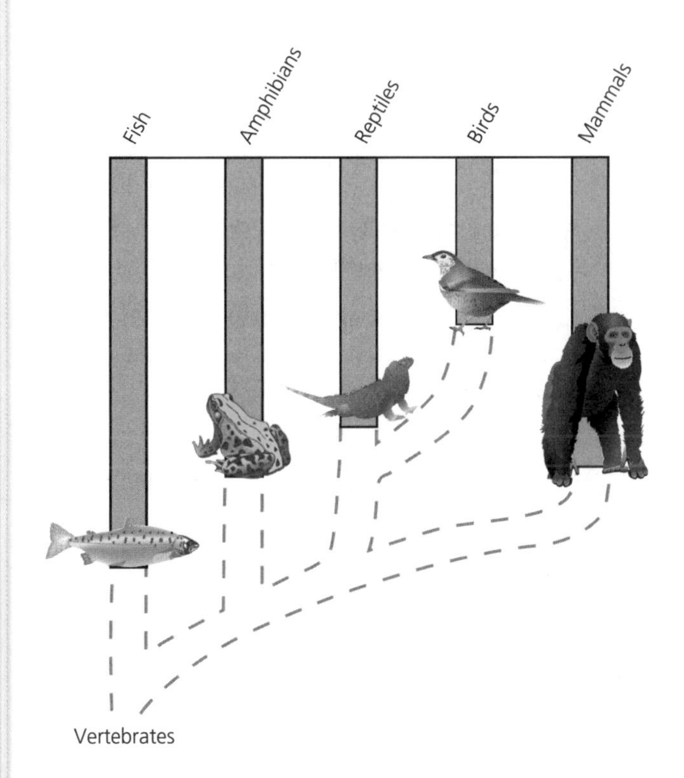

You need to be able to:
- **identify the differences between Darwin's theory of evolution and conflicting theories**
- **suggest reasons why Darwin's theory of natural selection was only gradually accepted**
- **suggest reasons for the different theories and why scientists cannot be certain about how life began on Earth.**

Darwin's Theory

Charles Darwin sailed around the world in the 1830s collecting evidence for his 'theory of natural selection'. This states that:

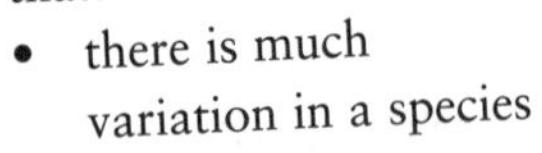

- there is much variation in a species

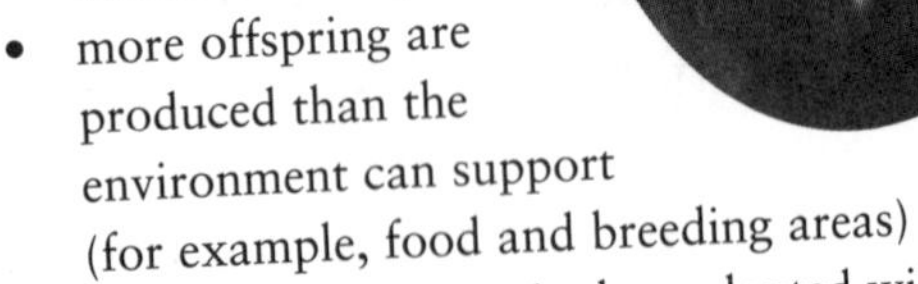

- more offspring are produced than the environment can support (for example, food and breeding areas)
- only the ones best suited or adapted will survive and breed. This is known as natural selection or 'survival of the fittest'
- those that survive pass their genes on to their offspring. Eventually the species which is less suited will become **extinct**.

Darwin's theory was only gradually accepted because:
- religion, which disagreed with evolution, had an important place in society
- it is difficult to prove
- many scientists didn't (and some still don't) accept the theory
- attempts to demonstrate evolution through tests have failed.

The Conflicting Theories

1. **Lamarck's Theory**
   Jean Baptiste Lamarck thought that living things changed throughout their lives and that these changes were passed on to their offspring. So he had the idea that giraffes grew long necks to reach the highest leaves on a tree and then passed this trait on to their offspring. He would have also believed that if a person learnt to play the piano, that person's children would be born able to play the piano. His was 'the theory of inheritance of acquired characteristics'.

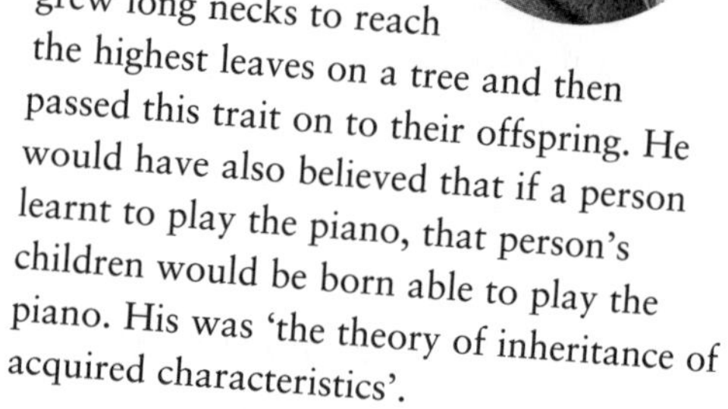

2. **Intelligent Design (ID)**
   This view states that certain structures within cells, such as DNA and mitochondria, are far too complicated to have evolved over time so must have been put there by some other higher being or creator.

Reasons for the different theories may include:
- **religion** – people who are religious may not accept scientific theories because they believe in a creator
- **culture** – people's backgrounds can influence the way they think
- **evidence** – certain theories may have more evidence to support them than others
- **knowledge** – people believe what they know
- **status of theorists / scientists** – people may be more likely to believe the ideas of renowned scientists or prominent people.

Remember that there are many different theories and scientists cannot be absolutely certain about how life began on Earth because it is difficult to find evidence to prove any theory.

Theories are based on the best evidence available at that time.

No-one experienced the beginning of life on Earth so it is impossible to ever be certain how it began. Even today, we are still finding out about things and developing our scientific knowledge.

You need to be able to interpret evidence relating to evolutionary theory.

Example

Fossil evidence has been discovered of what appears to be a 'sea monster', thought to be 135 million years old.

The animal's large skull was found in an area in Argentina that was once part of the Pacific Ocean.

Measuring four metres (13 feet) in length, the *Dakosaurus* had four paddle-like limbs, thought to be used for balance. It had a tail like a fish that propelled it through the sea, and a head that bore a resemblance to a carnivorous dinosaur.

The fossil displays a short, high snout and big, serrated teeth similar to a crocodile's.

Dakosaurus' serrated tooth

Every other known marine reptile had a long snout and sharp, identical teeth, which were used to deal with their main prey – small fish.

Dakosaurus's unusual snout, sharp, jagged teeth and huge jaws suggest that it hunted for sea reptiles and other large marine creatures, using its teeth to bite into and chop up its prey.

Question

Was Dakosaurus related to today's crocodiles?

Evidence

For crocodile:

- Large skull like a crocodile.
- Short, high snout similar to a crocodile.
- Serrated (jagged) teeth like a crocodile.

Against crocodile:

- Tail like a fish.
- Paddle-like limbs, no legs.
- Large jaws and teeth suggest it preyed on large sea reptiles and other sea creatures.

Interpretation

- *State your view*
  Dakosaurus was a sea monster related to today's crocodiles.
- *Explain your view*
  This seems to be true because the **fossils** were similar to crocodiles – they had a short, high snout and serrated teeth.
- *Conclude*
  The characteristics are not found in marine reptiles. This means that Dakosaurus was a reptile with many similarities to today's crocodiles. However, unlike crocodiles, Dakosaurus was adapted to live in the sea as a sea creature.

Practice Questions

…d two new drugs, 'Reduce' and 'LessU' which they claim will help people to lose … shows the results of some clinical trials.

…ken	Number of Volunteers in Trial	Average Amount of Weight Lost in 6 Weeks (kg)
Reduce	3250	3.2
LessU	720	5.8
Placebo	2800	2.6

a) What is a placebo? **(1 mark)**

b) Why was a placebo given to some volunteers? **(2 marks)**

c) In which trial would the data be most reliable? Explain your answer. **(2 marks)**

d) What conclusion could you draw from the data? **(2 marks)**

e) Suggest two things scientists would need to control to ensure the conclusions were valid. **(2 marks)**

f) State two ways the drugs could have been tested in the laboratory before being given to the volunteers. **(2 marks)**

g) Suggest two reasons why it is necessary to trial new drugs. **(2 marks)**

2 Gemma and Paul are twins.

a) Which of the following factors are they likely to have inherited? **(2 marks)**

hair colour **shape of their ear lobes** **weight** **spots**

b) Complete the following passage by filling in the spaces with words from the list. **(4 marks)**

glucose **nucleus** **dna** **genes** **protein**

cytoplasm **fat** **RNA** **23** **46**

Humans have .... pairs of chromosomes. Chromosomes are found in the .... of cells. They are made from .... and contain ....

3 **a)** When did scientists think simple life forms first appeared on Earth? **(1 mark)**

3 million years ago **30 million years ago** **3 billion years ago**

b) Darwin suggested a theory of evolution that many people were reluctant to accept. Give two reasons why people did not accept his theory. **(2 marks)**

c) Many organisms which once lived are now extinct. Give two reasons why animals become extinct. **(2 marks)**

C1.1 The fundamental ideas in chemistry

Atoms are the building blocks of chemistry. Atoms contain protons, neutrons and electrons. Elements are substances containing only one type of atom – when elements react they produce compounds. Elements are grouped together in the Periodic Table. To understand this, you need to know:

- what atoms, elements and compounds are
- what the subatomic particles in atoms are and how electrons are arranged
- how elements are arranged in the Periodic Table (including knowledge of alkali metals, halogens and noble gases)
- how compounds are formed.

Atoms

All substances are made of **atoms** (very small particles). Each atom has a small central **nucleus** made up of **protons** and **neutrons** that is surrounded by **electrons**.

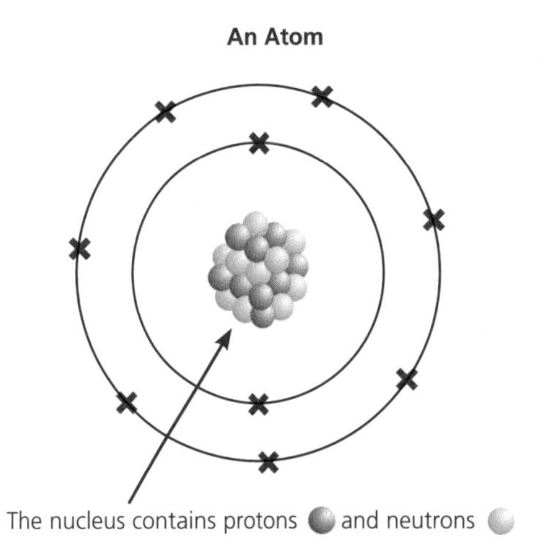

Elements

A substance that contains only one sort of atom is called an **element**. There are over 100 different elements.

The atoms of each element are represented by a different chemical symbol, for example, O for oxygen, Na for sodium, C for carbon, and Fe for iron. Elements are arranged in the Periodic Table (see below). The groups in the Periodic Table contain elements that have similar properties.

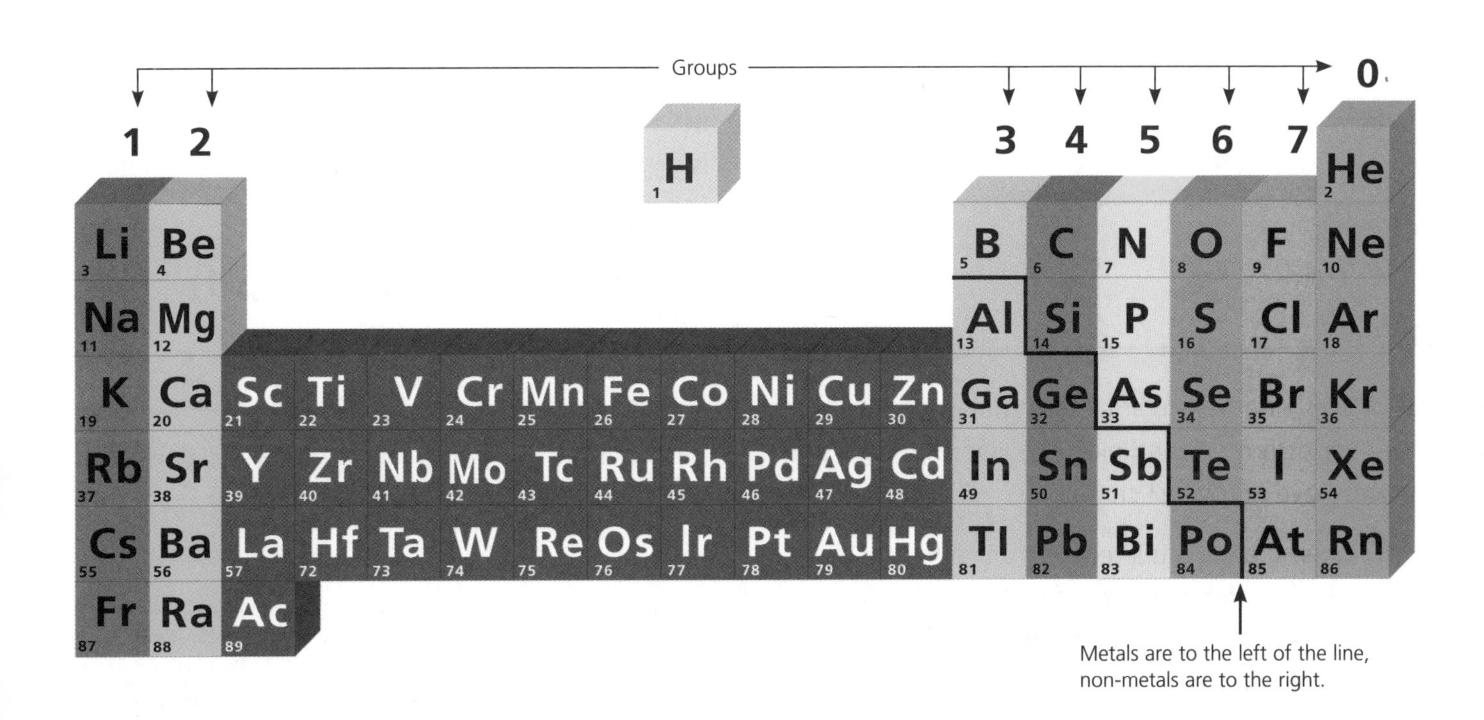

Subatomic Particles

The diagram below shows the subatomic particles in an atom of helium.

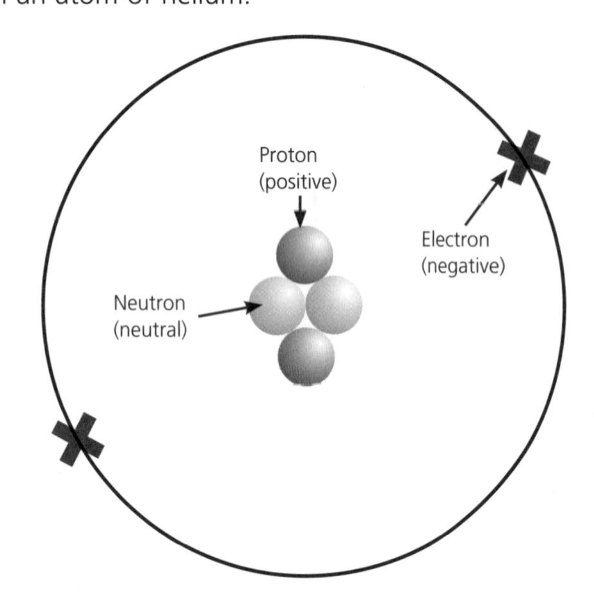

Atoms have a small central nucleus that is made up of protons and neutrons. The nucleus is surrounded by electrons. Protons, neutrons and electrons have relative electrical charges.

Atomic Particle	Relative Charge
Proton	+1
Neutron	0
Electron	–1

All atoms of a particular element have an equal number of protons and electrons, which means that atoms have no overall charge.

All atoms of a particular element have the same number of protons. Atoms of different elements have different numbers of protons. This is known as the **atomic number**.

Elements are arranged in the modern Periodic Table in order of increasing atomic number.

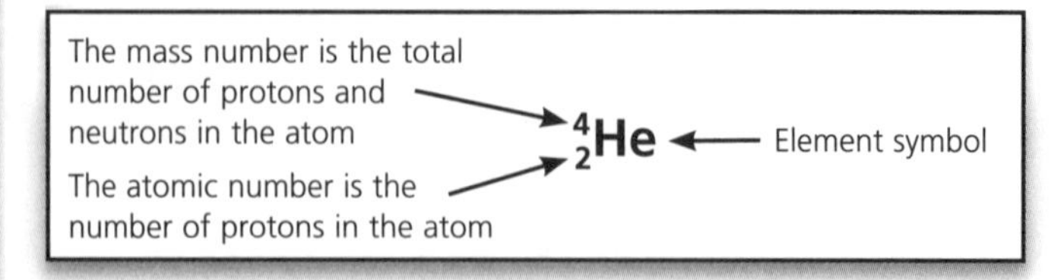

Electron configuration tells us how the electrons are arranged around the nucleus in energy levels or shells.

The electrons in an atom occupy the lowest available energy levels (i.e. the innermost available shells).

- The first level or shell can hold a maximum of 2 electrons.
- The energy levels or shells after this can hold a maximum of 8 electrons.
- The energy of the shells increases as they get further from the nucleus.

We write electron configurations as a series of numbers, e.g. oxygen is 2,6 and aluminium is 2,8,3.

The Periodic Table arranges the elements in terms of their electronic structure. Elements in the same group (column) have the same number of electrons in their outer shell (this number is the same as the group number). Elements in the same group therefore have similar properties. From left to right, across each period (row), a particular energy level is gradually filled with electrons. In the next period, the next energy level is filled, etc.

The Alkali Metals (Group 1)

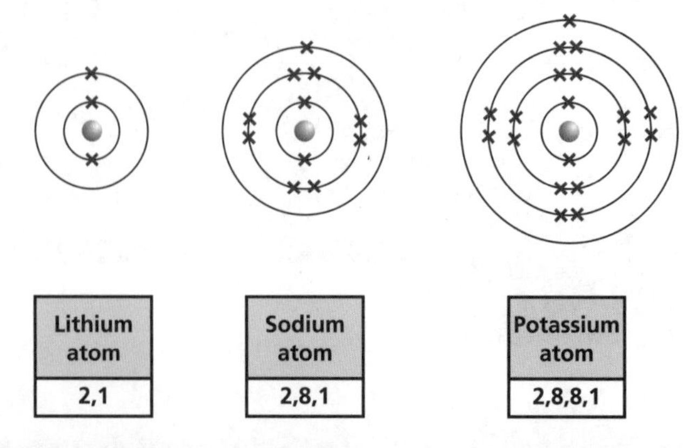

The alkali metals all have similar properties because their atoms have the same number of electrons (one) in their outermost shell, i.e. the highest occupied energy level contains one electron. They react with non-metal elements to form ionic compounds, in which the metal ions have a single positive charge.

Because the alkali metals all have one electron in their outer shell, they react in a similar way to each other. All the alkali metals react quite vigorously when you add them to water. They all produce an alkaline solution (a metal hydroxide) and hydrogen gas. For example:

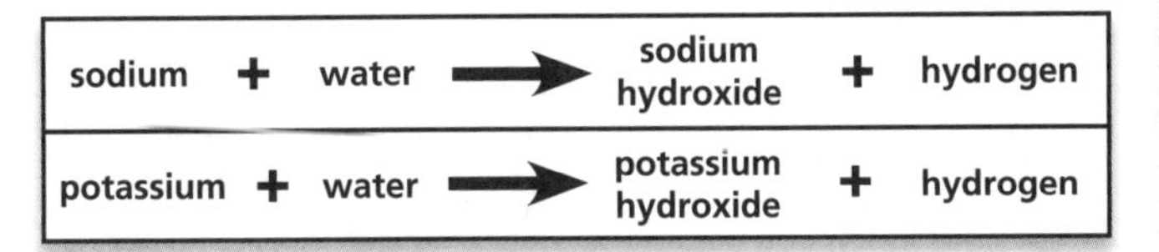

sodium	+	water	→	sodium hydroxide	+	hydrogen
potassium	+	water	→	potassium hydroxide	+	hydrogen

The alkali metals will all react with oxygen to produce a metal oxide. For example:

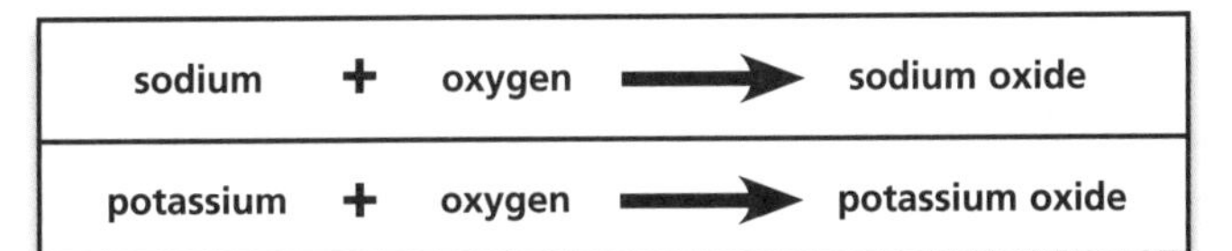

sodium	+	oxygen	→	sodium oxide
potassium	+	oxygen	→	potassium oxide

The Halogens (Group 7)

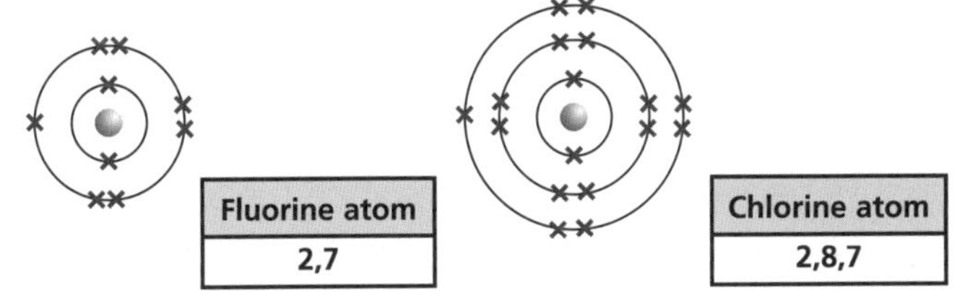

The halogens all have similar properties because their atoms have the same number of electrons (7) in their outermost shell, i.e. the highest occupied energy level contains 7 electrons.

They react with alkali metals to form ionic compounds, in which the halide ions have a single negative charge.

Noble Gases (Group 0)

The noble gases are inert (unreactive) gases. They are unreactive because they have a stable electron arrangement. The outer shell of each noble gas atom is full. Helium has 2 outer electrons. All other noble gases have 8 outer electrons.

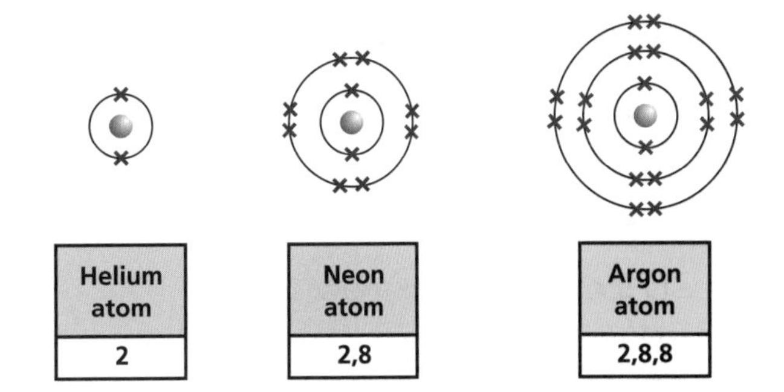

Mixtures and Compounds

A **mixture** consists of two or more elements or compounds that are not chemically combined. The properties of the substances remain unchanged and specific to each substance.

Compounds are substances in which the atoms of two or more elements are chemically combined (not just mixed together). The atoms are held together by chemical bonds.

Atoms can form chemical bonds by:
- sharing electrons (covalent bonds)
- gaining or losing electrons (ionic bonds).

Compounds formed from the reaction of metal atoms and non-metal atoms consist of ions. The metal atoms lose electrons to form positive ions. For example, a sodium atom (Na) loses an electron to become a sodium ion (Na^+). The non-metal atoms gain electrons to form negative ions. For example, a chlorine atom (Cl) gains an electron to form a chloride ion (Cl^-). These compounds are held together by ionic bonds, e.g. sodium chloride (NaCl).

Compounds formed from only non-metal atoms consist of molecules. In these, atoms share electrons and are held together by covalent bonds, e.g. carbon dioxide (CO_2).

Either way, when atoms form chemical bonds the arrangement of the outermost shell of electrons changes resulting in each atom getting a complete outer shell of electrons.

For most atoms this is eight electrons but for hydrogen it is only two.

Chemical Formulae

Compounds are represented by a combination of numbers and chemical symbols called a **formula**, e.g. ZnO or $2H_2SO_4$.

Chemists use formulae to show:
- the different elements in a compound
- the ratio of atoms of each element in the compound.

In chemical formulae, the position of the numbers tells you what is multiplied. Smaller numbers that sit below the line (subscripts) only multiply the symbol that comes immediately before it, and large numbers that are the same size as the letters multiply all the symbols that come after.

For example:
- H_2O means (2 × H) + (1 × O)
- 2NaOH means 2 × (NaOH) **or** (2 × Na) + (2 × O) + (2 × H)
- $Ca(OH)_2$ means (1 × Ca) + (2 × O) + (2 × H).

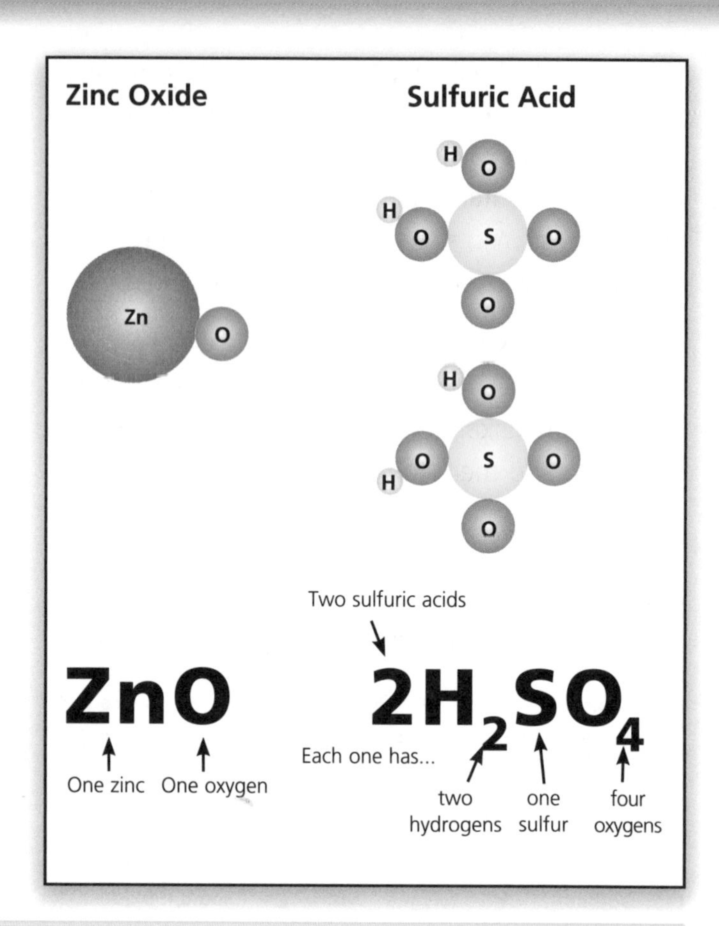

Chemical Reactions

You can show what has happened during a reaction by writing a **word equation** showing the substances that react (the **reactants**) on one side of the equation and the new substances formed (the **products**) on the other.

The total mass of the products of a chemical reaction is always equal to the total mass of the reactants.

This is because the products of a chemical reaction are made up from exactly the same atoms as the reactants – no atoms are lost or made!

That means that chemical symbol equations must always be balanced – there must be the same number of atoms of each element on the reactant side of the equation as there is on the product side.

Example

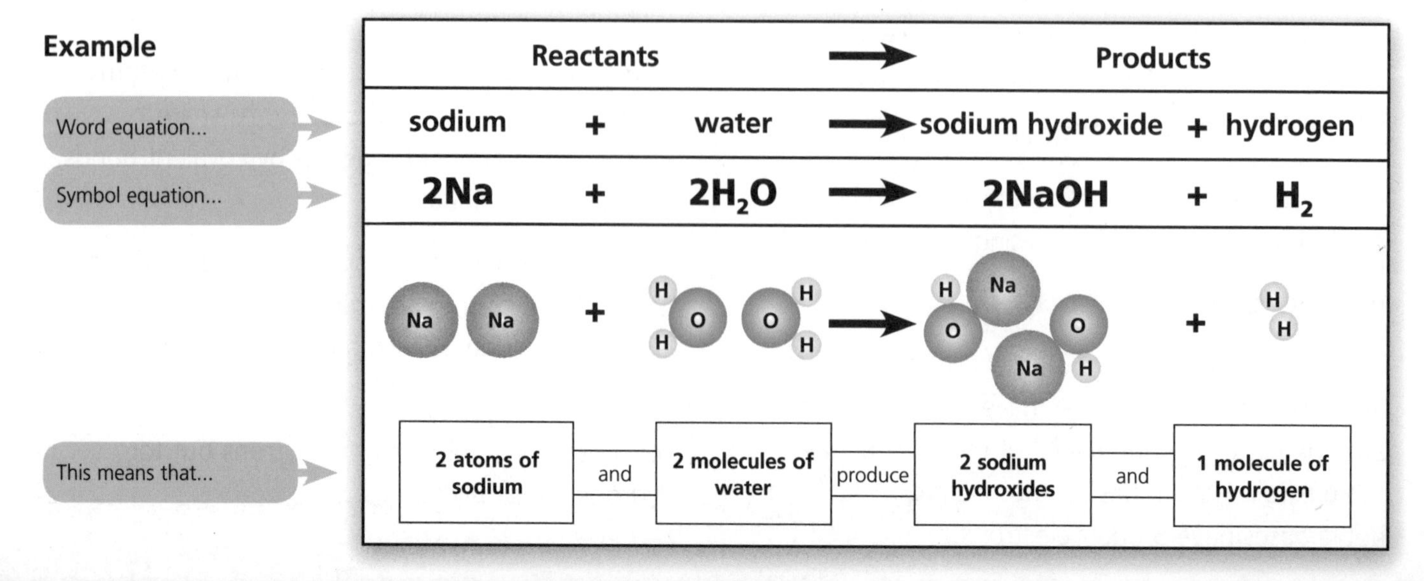

HT Writing Balanced Equations

Follow these steps to write a balanced equation:

1. Write a word equation for the chemical reaction.
2. Substitute formulae for the elements or compounds involved.
3. Balance the equation by writing numbers in front of the reactants and / or products.
4. Write the balanced symbol equation.

Example 1 – The reaction between magnesium and oxygen.

	Reactants			→	Products
1 Write a word equation	magnesium	+	oxygen	→	magnesium oxide
2 Substitute formulae	Mg	+	O_2	→	MgO
3 Balance the equation	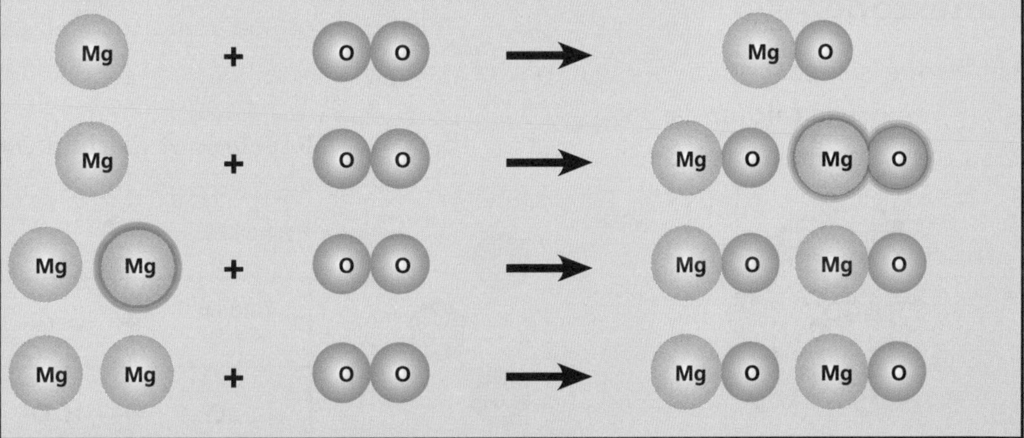 Mg	+	O O	→	Mg O
	Mg	+	O O	→	Mg O Mg O
	Mg Mg	+	O O	→	Mg O Mg O
	Mg Mg	+	O O	→	Mg O Mg O
4 Write a balanced symbol equation	2Mg	+	O_2	→	2MgO

- There are two **O**s on the reactant side, but only one **O** on the product side. We need to add another **MgO** to the product side to balance the **O**s
- We now need to add another **Mg** on the reactant side to balance the **Mg**s
- There are two magnesium atoms and two oxygen atoms on each side – **it is balanced**.

Example 2 – The production of ammonia.

	Reactants			→	Products
1 Write a word equation	nitrogen	+	hydrogen	→	ammonia
2 Substitute formulae	N_2	+	H_2	→	NH_3
3 Balance the equation	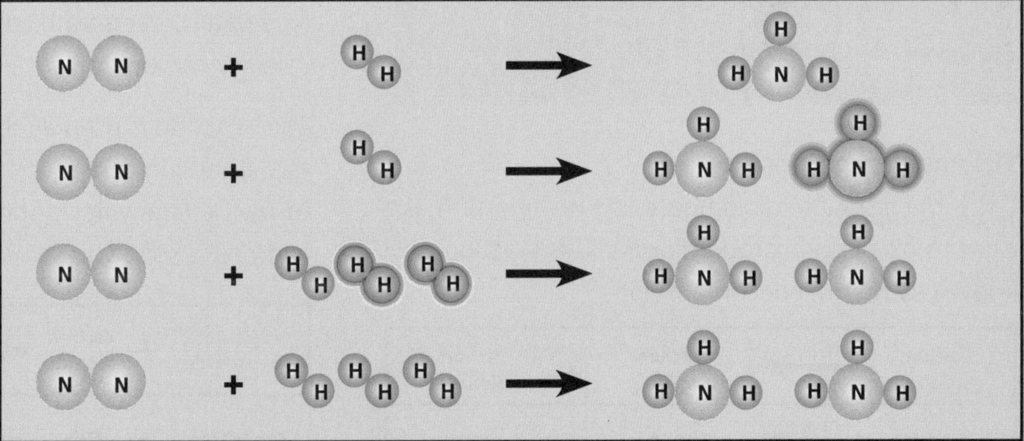 N N	+	H H	→	H H N H
	N N	+	H H	→	H H N H H H N H
	N N	+	H H H H H H	→	H H N H H H N H
	N N	+	H H H H H H	→	H H N H H H N H
4 Write a balanced symbol equation	N_2	+	$3H_2$	→	$2NH_3$

- There are two **N**s on the reactant side, but only one **N** on the product side. We need to add another **NH_3** to the product side to balance the **N**s
- We now need to add two more **H_2**s on the reactant side to balance the **H_2**s
- There are two nitrogen atoms and six hydrogen atoms on each side – **it is balanced**.

C1.2 Limestone and building materials

Rocks provide essential building materials. Limestone is a naturally occurring resource that provides a starting point for the manufacture of cement and concrete. To understand this, you need to know:

- why limestone is a useful resource
- how limestone is used to produce building materials.

Limestone

Limestone is a sedimentary rock that consists mainly of the compound **calcium carbonate** ($CaCO_3$). It is cheap, easy to obtain and has many uses.

Building Material

Limestone can be **quarried** and cut into blocks, and used to build walls of houses in regions where limestone is plentiful.

Thermal Decomposition

Calcium carbonate decomposes on heating to make calcium oxide and carbon dioxide. This reaction is known as thermal decomposition.

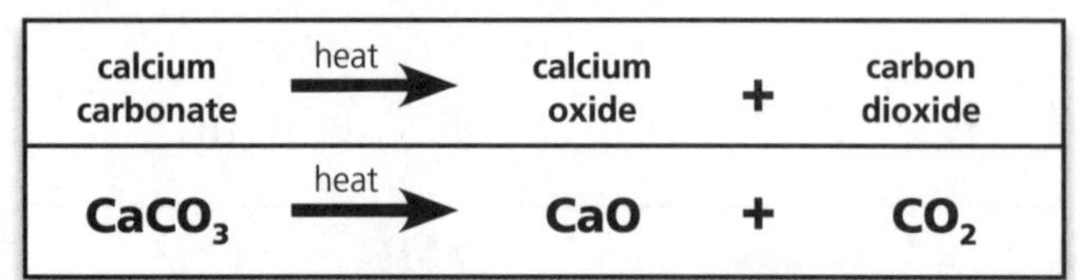

calcium carbonate	heat →	calcium oxide	+	carbon dioxide
$CaCO_3$	heat →	CaO	+	CO_2

Other metal carbonates undergo thermal decomposition on heating. They also produce a metal oxide and carbon dioxide gas. For example:

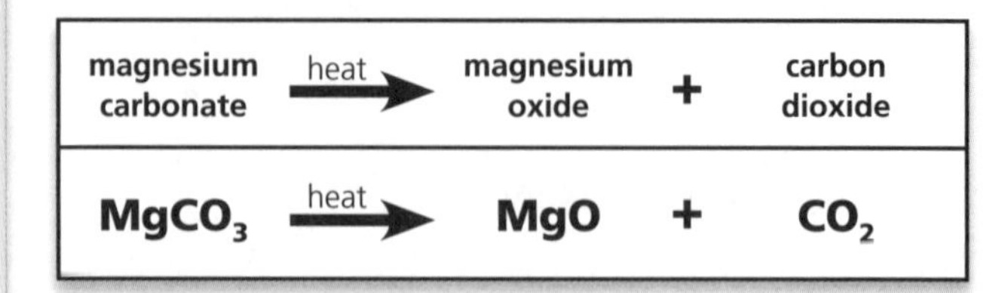

magnesium carbonate	heat →	magnesium oxide	+	carbon dioxide
$MgCO_3$	heat →	MgO	+	CO_2

N.B. The carbonates of other metals behave very similarly when they are heated.

Some Group 1 metal carbonates require higher temperatures than a Bunsen burner can provide in order to undergo thermal decomposition.

Reaction of Calcium Oxide

Calcium oxide reacts with water to form calcium hydroxide.

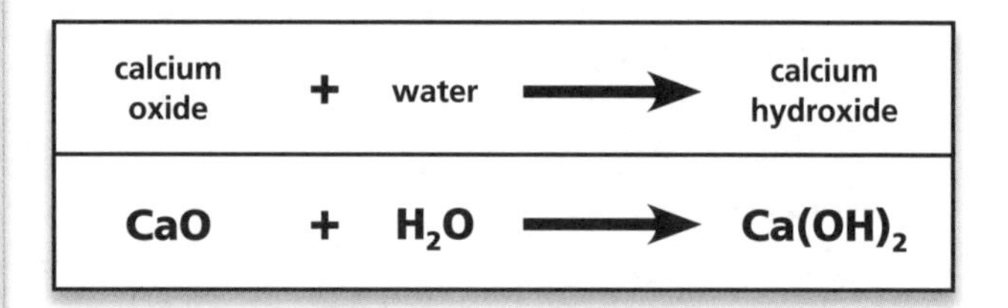

calcium oxide	+	water	→	calcium hydroxide
CaO	+	H_2O	→	$Ca(OH)_2$

Calcium hydroxide (like all metal hydroxides), is a strong alkali. It can be used to neutralise soils and lakes much faster than powdered limestone.

Reaction of Calcium Hydroxide

Calcium hydroxide solution (also known as limewater) reacts with carbon dioxide to form calcium carbonate.

This reaction is used as the test for carbon dioxide gas. Bubbling carbon dioxide gas through limewater turns the limewater cloudy (this is actually calcium carbonate solid you see forming).

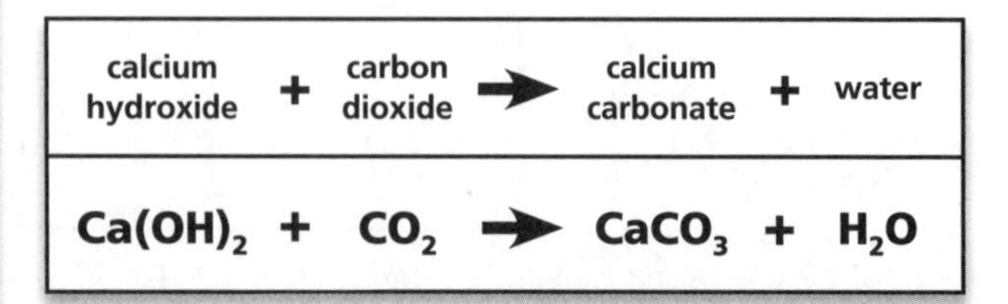

calcium hydroxide	+	carbon dioxide	→	calcium carbonate	+	water
$Ca(OH)_2$	+	CO_2	→	$CaCO_3$	+	H_2O

Limestone Cycle Summary

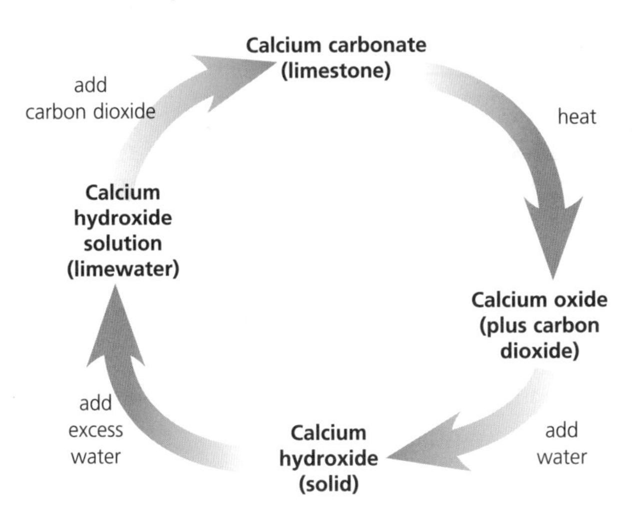

Cement, Mortar and Concrete

Powdered limestone and powdered clay are roasted in a rotary kiln to produce dry **cement**.

When the cement is mixed with sand and water it produces **mortar**, which is used to hold together bricks and stone during building.

When the cement is mixed with water, sand and gravel (crushed rock) a slow reaction takes place in which a hard, stone-like building material, called **concrete**, is produced.

Reaction of Metal Carbonates

Carbonates such as calcium carbonate (limestone) can react with acids to form a salt, water and carbon dioxide gas. All metal carbonates that you will meet in this unit behave in the same way.

metal carbonate + acid → salt + water + carbon dioxide
calcium carbonate + hydrochloric acid → calcium chloride + water + carbon dioxide
$CaCO_3 + 2HCl \rightarrow CaCl_2 + H_2O + CO_2$

This means that, over time, limestone (calcium carbonate) can become damaged by acid rain.

magnesium carbonate + hydrochloric acid → magnesium chloride + water + carbon dioxide
$MgCO_3 + 2HCl \rightarrow MgCl_2 + H_2O + CO_2$

You need to be able to consider and evaluate the environmental, social and economic effects of exploiting limestone and producing building materials from it.

Advantages	Disadvantages
• Limestone is found naturally, so can be quarried relatively easily. • Using local stone to build new houses makes them 'fit in' with older houses. • Better roads will be built to cope with quarry traffic. • Creates more jobs locally. • Other industries (e.g. cement makers) will be attracted to the area, providing more job opportunities. • The quarry might invest in the local community in a bid to 'win over' the locals.	• Could be more expensive to quarry limestone than to use another building material. • Quarries destroy the landscape and the habitats of animals and birds. • Increased traffic to and from the quarries. • Noise pollution. • Health problems arising from the dust particles, e.g. asthma. • Reduced tourism in the area.

You need to be able to evaluate the advantages and disadvantages of using limestone, concrete and glass as building materials.

Material	Advantages	Disadvantages
Limestone	• Widely available. • Easy to cut. • Cheaper than many other building materials, e.g. marble. • Can be used to produce cement, concrete and glass.	• Susceptible to acid rain – the dilute acid dissolves the limestone very slowly, wearing it away.
Concrete	• Can be moulded into different shapes, e.g. panels and blocks which can be put together easily in buildings. • Quick and cheap way to construct buildings. • Does not corrode, so is a good alternative to metal. • Can be reinforced using steel bars so that it is safer and has a wider range of uses.	• Low tensile strength* and can crack and become dangerous, especially in high-rise buildings. • Looks unattractive.
Glass	• Transparent, therefore useful for windows and parts of buildings where natural light is wanted. • Can be toughened or made into safety glass. N.B. You don't need to know about glass in detail – but be aware that you may be given sufficient information to make comparisons with other materials	• Breaks easily. • Not always the cheapest or safest option. • Toughened and safety glass are expensive.

**Tensile strength refers to a material's ability to resist breaking when under tension.*

All of these materials are resistant to fire and rot, and for most building requirements are strong enough to resist attack from animals and insects, which make them a better choice than wood. However, there may be cheaper, safer and more aesthetically pleasing materials that are also suitable for the job.

C1.3 Metals and their uses

We use metals constantly in our everyday lives. Most metals are obtained from ores, which are naturally occurring rocks that provide an economic starting point for the manufacture of metals. Copper, iron, aluminium and titanium are examples of commonly used metals extracted from ores. Using more of these metals is creating a shortage of some ores, e.g. copper-rich ores, so new methods of extraction are being researched. To understand this, you need to know:

- what an ore is
- how we obtain different metals from their ores
- the properties and uses of different metals such as copper, iron, aluminium and titanium
- what an alloy is and why they are important
- the impact on the environment caused by the extraction of metals and methods being used to overcome this.

Ores

The Earth's crust contains many naturally occurring elements and compounds called **minerals**. A metal **ore** is a mineral or mixture of minerals from which economically viable amounts of pure metal can be **extracted**. This can change over time. After ores have been mined they may be concentrated before the metal is extracted and purified.

Extracting Metals from their Ores

The method of extraction depends on how reactive the metal is. Unreactive metals like gold exist as the metal itself (native metal). They are obtained through physical processes such as panning.

Most metals are found as **metal oxides** or compounds that can be easily changed into a metal oxide. To extract a metal from its oxide the oxygen must be removed by heating the oxide with another element in a chemical reaction. This process is called **reduction**.

Metals that are less reactive than carbon can be extracted from their oxides by heating them with carbon. (The carbon is a more reactive element, so it will displace the metal and form a compound with the oxygen.)

Metals that are more reactive than carbon are extracted from their ores using **electrolysis**. The ore is heated until it is liquid (molten) before the metal can be extracted.

Electrolysis requires large amounts of electricity and is, therefore, an expensive method to use. Reactive metals such as aluminium have to be extracted using electrolysis.

The Transition Metals

In the centre of the **Periodic Table**, between Group 2 and Group 3, is a block of metallic elements called the **transition metals**. These include iron, copper, platinum, mercury, chromium, titanium and zinc.

These metals are **hard** and mechanically **strong**. They have **high melting points** (except mercury – which is liquid at room temperature).

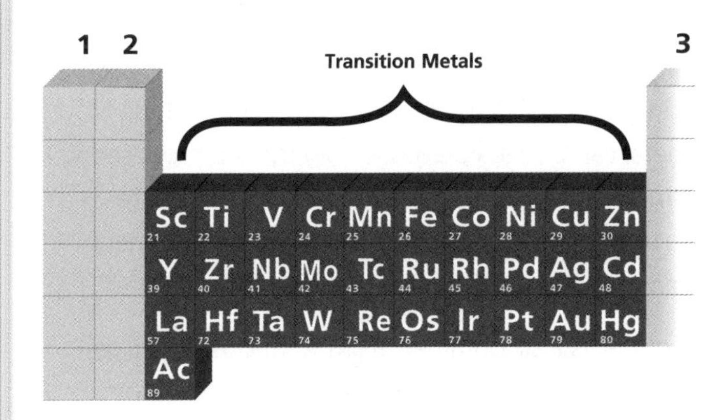

Transition metals, like all other metals, are **good conductors** of heat and electricity, and can also be easily bent or hammered into shape.

These properties make transition metals very useful as structural materials, and as electrical and thermal conductors.

Metals and their Uses

Copper

Copper is used for electrical wiring and plumbing because of its properties. Copper:

- is a good conductor of heat and electricity (good for use as electrical wires)
- is malleable (can be hammered into shape and bent), but is also hard enough to make pipes
- does not react with water (good for use as water pipes)
- can be drawn into wires (ductile).

Copper is a very valuable metal and can be extracted from copper-rich ores using a furnace. This process is known as smelting. The resultant copper is then purified using electrolysis. Because copper is such a versatile and useful material we are rapidly depleting our supplies. Therefore, the supply of copper-rich ores is limited.

To enable us to continue using copper, new ways of extracting the metal from low-grade ores are currently being researched. These new extraction methods include phytomining and bioleaching, and not only allow us to use low-grade ores but they also limit the environmental impact of more traditional mining methods.

Phytomining uses plants to absorb metal compounds. The plants are then burned and the metal can be separated out from the ash produced. **Bioleaching** uses bacteria to produce **leachate** solutions that contain metal compounds. The metal can then be extracted from this solution.

Copper is also obtained from solutions of copper salts using electrolysis or displacement reactions using the more reactive metal iron. Electrolysis involves the positive copper ions being attracted to the negative electrode.

Aluminium and Titanium

Aluminium and titanium are both useful metals because they have a low density and are resistant to corrosion. Aluminium reacts with oxygen from the air and so a layer of aluminium oxide coats the metal, which prevents further corrosion. Aluminium is used in making drink cans, window frames and aeroplanes. Titanium is used in making aeroplanes, nuclear reactors and replacement hip joints.

Aluminium and titanium are more reactive than carbon and, therefore, cannot be extracted from their oxides using reduction. Aluminium is extracted using **electrolysis**, which makes it expensive.

Electrolysis is very complex – there are lots of different stages – and requires a large amount of energy. This makes it very expensive. So, we should recycle metals wherever possible to:

- save money and energy
- make sure we do not use up all the natural resources
- reduce the amount of mining because it is damaging the environment.

Iron

Iron is less reactive than carbon and can, therefore, be extracted from its ore by reduction using carbon in a blast furnace.

Molten iron obtained from a blast furnace contains roughly 96% iron along with 4% carbon and other metals. Because it is impure, the iron is very brittle and has limited uses. To produce **pure iron**, all the impurities would have to be removed.

Alloys

An alloy is a mixture of metals, or a metal and at least one other element. The added element disturbs the regular arrangement of the metal atoms so that the layers do not slide over each other so easily. Alloys are, therefore, usually stronger and harder than pure metals. Many of the metals we use everyday are alloys.

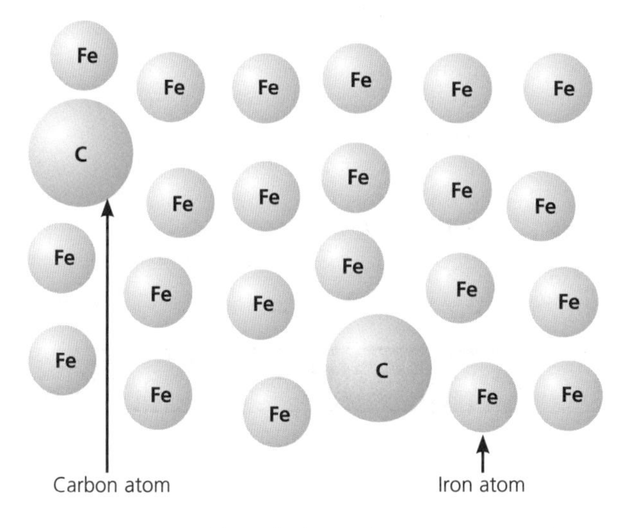

Pure copper, gold and aluminium are too soft for many uses. They are mixed with small amounts of similar metals to make them harder for everyday use.

Steel

Pure iron is soft, and iron from the blast furnace is brittle and easily corrodes (rusts), so most iron is converted into steel. Carbon is usually added to iron to make the steel (an alloy). However, some steels, such as stainless steel, are produced by adding other metals to the iron.

Most iron is converted into steel because this makes it more useful.

Alloys like steel are developed to have the necessary properties for a specific purpose. In steel, the amount of carbon and/or other elements determines its properties:

- steel with a high carbon content is hard and strong, e.g. for screwdrivers
- steel with a low carbon content is soft and easily shaped – mild steel (0.25% carbon) is easily pressed into shape, e.g. for cars
- steel that contains chromium and nickel is called stainless steel – it is hard and resistant to corrosion, e.g. for knives and forks.

Recycling and Environmental Impact

Metal ores are limited resources. The continued extraction and use of metals means that we are running out of some metals and, therefore, we need to recycle the metals we do have in order to preserve these limited resources. Extraction often requires the use of large amounts of energy, which is both expensive and damaging to the environment.

Extraction methods often lead to pollution or contamination of land by metals such as cadmium, nickel and cobalt. Plants like cabbages (brassicas) can be used to remove unwanted metals from the soil. The metals are then removed by taking the plants out of the soil.

Smart Alloys

Smart alloys belong to a group of materials that are being developed to meet the demands of modern engineering and manufacturing. These materials respond to changes in their environment, e.g. temperature, moisture, pH and electrical and magnetic fields.

Smart alloys (also called shape-memory alloys) remember their shape. They can be deformed, but will return to their original shape (usually when they are heated).

You need to be able to consider and evaluate the social, economic and environmental impacts of exploiting metal ores, of using metals and of recycling metals.

Example

The Lonsdale News, Tuesday 20 September 2011 17

Local Village Launches National Campaign

The villagers of Littlehampton are speaking out to show us all how we can help to reduce the damage being done to our environment.

A large metal extraction plant was built near Littlehampton 15 years ago and the village has suffered from the effects of the industry ever since. The plant has had a detrimental impact not only on the look of the area but also on its environment. The pollution has made its way into rivers and streams, and the local wildlife group reports that certain species are dwindling in number.

These bad effects are not limited to wildlife. The noise that comes from the factory is annoying, but more worrying is the dramatic increase in the number of local people who suffer from asthma as a result of the dust particles discharged into the air.

Now, after proposals to extend the plant have been revealed, villagers have joined together to recycle metals and want to encourage others to do the same. Campaign spokesman Bob Jeffries, 42, says, 'We know that metals are useful materials and that extraction needs to be done, but we hope to encourage people to recycle what they can. This should reduce the demand for newly extracted metals and remove the need for new plants. Not only will we be improving our standard of living but we will also be helping to reduce the pressure and impact on our environment'.

Recycling is a much better option because it uses less energy, which makes it a much cheaper process. The more times a material is recycled, the more cost-effective it becomes.

Councillors listened to the views of local residents and agreed to implement a recycling scheme. Five large recycling bins have been brought in to the car park at the local supermarket. Said councillor Cilla Jackson, 56, 'The scheme has had a much better response than we had hoped for; I just hope the enthusiasm for it is maintained.'

For information on how you can do your bit go to www.recyclenow.com

Method	Advantages	Disadvantages
Extracting metals	• Provides jobs and income locally. • Provides raw materials for industry. • Local facilities (e.g. roads) improved to cope with additional traffic.	• Destroys the landscape. • Leads to a reduction in tourism. • Noise and dust pollution. • Traffic problems.
Recycling metals	• Saves energy (e.g. less energy used to recycle aluminium than to electrolyse the ore it comes from). • Less pollution produced because fewer materials are sent to landfill sites. • Less pressure placed on the environment (the more material that is recycled, the less pressure there is to find new materials to mine, extract, etc.).	• Individual apathy. • Availability and collection of recycling facilities can make recycling difficult.

You need to be able to evaluate the advantages and disadvantages of using metals as structural materials and as smart materials.

Use of Metal	Advantages	Disadvantages
Structural material	• Hard, tough and strong. • Do not corrode easily. • Can be bent or hammered into shape. • Alloys of metals are harder than pure metals. • Relatively inexpensive at present.	• Iron is naturally very soft so needs to be mixed with other metals to form steel. • Conduct electricity and heat – this might not be what is wanted. • Some metals, particularly iron, can be corroded by water and other chemicals – this weakens and eventually wears away the metal. • The supply of metal ores from the Earth's crust is decreasing as more is extracted – eventually the supplies will run out.
Smart material	• Can be produced by mixing metals – have many advantageous properties over the original metals. • Good mechanical properties, e.g. strong and resist corrosion. • Can return to their original shape when heated (known as the shape-memory effect) – used in thermostats, coffee pots, hydraulic fittings. • More bendy than normal metals, therefore harder to damage. • Have new properties, such as pseudo-elasticity, which can be exploited in diverse ways such as in glasses frames, bra underwires and orthodontic arches. • Can be changed by passing an electrical current or a magnetic field through them or by heating. • Not much temperature change required (sometimes as little as 10°C) to change the structure.	• Expensive to manufacture. • Fatigue easily – a steel component can survive for around 100 times longer than a smart material under the same pressure.

C1.4 Crude oil and fuels

Crude oil is found trapped in porous rocks and consists of many different hydrocarbons. Crude oil can be separated by fractional distillation into different fractions, some of which are very useful as fuels. Other useful fuels can be produced from plant material (known as biofuels). Crude oil is a non-renewable resource whereas biofuels are a renewable resource. To understand this, you need to know:

- the difference between a compound and a mixture
- how crude oil can be fractionally distilled to produce fuels
- that most fuels contain carbon and hydrogen, and sometimes sulfur
- what is produced when fuels burn and the impact that this has on the environment
- what alkanes are, their general formula and how to represent them.

Crude Oil

Crude oil is a mixture containing many different compounds. Most of these compounds are **hydrocarbons** (they contain only the atoms carbon and hydrogen). Mixtures contain different elements or compounds that are not chemically combined and, therefore, each part of the mixture keeps its own chemical properties.

Because crude oil is a mixture we can separate it using physical methods such as distillation. This separates crude oil into different fractions. Each fraction is made up of similar size hydrocarbons that have similar boiling points.

To separate crude oil using fractional distillation, the oil is vaporised (by heating), pumped into a fractionating column and the vapour is allowed to condense at different temperatures. The fractions, each of which contains hydrocarbons with a similar number of carbon atoms, are collected.

The smaller hydrocarbons (those with fewer carbon and hydrogen atoms) are more useful as fuels. This is because the properties of a hydrocarbon depend on the size of the carbon chain. The longer the carbon chain:

- the higher the boiling point
- the more viscous it is (i.e. it flows less easily)
- the less volatile it is (i.e. it does not vaporise/turn into a gas easily)
- the less flammable it is (i.e. it is much harder to ignite)
- the more soot it produces when it burns.

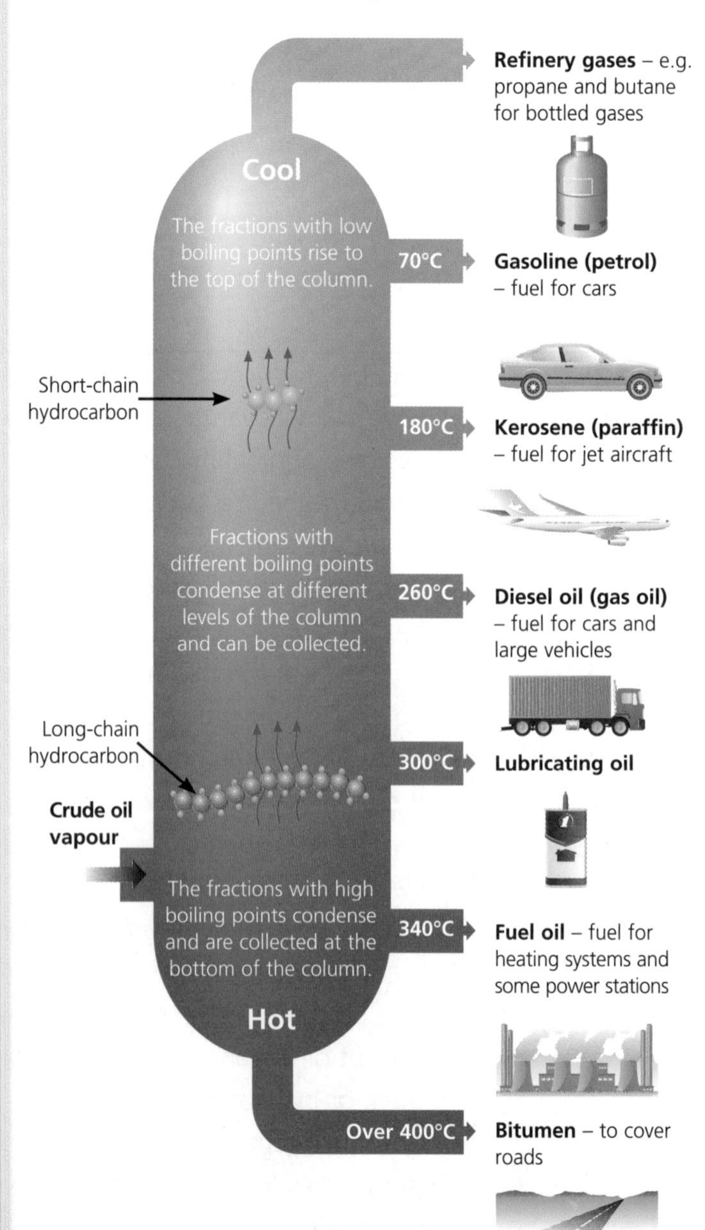

Alkanes

Most of the hydrocarbons in crude oil are saturated molecules called alkanes. Each of the carbons in an alkane molecule has four single bonds to either other carbon atoms or hydrogen atoms. All of the carbon–carbon bonds are single covalent bonds so we say the hydrocarbon is **saturated**.

Alkanes have the general formula C_nH_{2n+2}. Alkanes can be represented by structural formula and displayed formula. The table shows the first three alkanes in the series.

Name	Formula	Displayed Formula
Methane	CH_4	H–C(H)(H)–H
Ethane	C_2H_6	H–C(H)(H)–C(H)(H)–H
Propane	C_3H_8	H–C(H)(H)–C(H)(H)–C(H)(H)–H

Butane has 4 carbon atoms (C_4H_{10}).

Covalent bonds are shown in the above structures using a straight line (–).

Hydrocarbon Fuels

Fuels usually contain carbon and often hydrogen, and when burned they produce energy (in the form of heat). This is known as a **combustion reaction**.

Combustion reactions are actually oxidation reactions because the carbon is oxidised (gains oxygen) to form carbon dioxide and the hydrogen is oxidised to form water vapour.

Combustion reactions can produce different gases depending on whether the reaction is complete or partial. This depends on the amount of oxygen.

Short-chain hydrocarbons are very useful as fuels. When burned, hydrocarbon fuels release gases into the atmosphere. The gases include carbon dioxide, water vapour and possibly carbon monoxide. Sulfur is often present in the fuel as an impurity and so sulfur dioxide is often produced. Burning fuels can also lead to the production of nitrogen oxides and carbon particles (soot). Nitrogen oxides are formed at high temperatures.

Complete combustion of carbon produces carbon dioxide (CO_2) whereas partial or incomplete combustion produces carbon monoxide (CO) and solid particles that may contain unburned hydrocarbons and carbon particles (soot).

The gases and solid particles released by the combustion of fuels cause environmental problems:

- Carbon dioxide (CO_2) causes 'global warming' due to the greenhouse effect.
- Sulfur dioxide (SO_2) causes acid rain. This problem can be reduced by removing sulfur from the fuel before burning, or SO_2 from the waste gases after burning – however this can be costly. Power stations remove SO_2 from the waste gases produced when combustion occurs to reduce the amount of pollution they emit.
- Carbon particles (soot) cause global dimming (a reduction in the amount of sunlight reaching the Earth's surface).
- Nitrogen oxides can also cause acid rain.

Biofuels

Biofuels are produced from plant materials and include ethanol and biodiesel. There are advantages and disadvantages in using these types of fuels rather than fuels produced from crude oil.

You need to be able to consider and evaluate the social, economic and environmental impacts of the uses of fuels.

Example

Forget Fossil Fuels...

When hydrocarbons are burned they release harmful waste gases into the air. Burning fossil fuels has a considerable impact on the environment and its inhabitants. The solution is to use less fossil fuel by using alternative energy resources, using existing resources more efficiently and by making changes to lifestyles, such as car-sharing to reduce fuel consumption, etc.

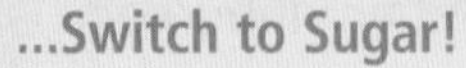

...Switch to Sugar!

Brazilian motorists have been converting their petrol-guzzling cars so they can be powered by ethanol. Ethanol, better known as grain alcohol, is easily distilled from sugar cane and is a cheap alternative to petrol. This new use for sugar cane has greatly affected farmers – never has there been such a high demand for sugar! This in turn has helped to push up the price of sugar to an all-time high. Scientists hope that the use of sugar cane as a viable alternative to petrol will grow, because it burns more cleanly than usual fuels and produces less of the harmful gas, carbon monoxide. However, it is not all good news. Alcohol releases less energy than petrol when it burns and it can be a health risk to filling station attendants.

You need to be able to evaluate developments in the production and uses of better fuels, for example, ethanol and hydrogen.

Example

Is Rocket Fuel the Way Forward?

Experts are carrying out research to find out if hydrogen gas, which is currently used as rocket fuel, could be an answer to our pollution problem. Hydrogen can be produced by passing an electric current through water, and when it burns it releases a lot of energy. Unlike other fuels which produce harmful gases when burned, hydrogen produces only water vapour, which does not pollute the atmosphere. A major consideration is the cost involved because the production of hydrogen requires a lot of electricity. It could also be quite risky because hydrogen burns explosively so it needs to be stored under special conditions.

Fuel	Advantages	Disadvantages
Fossil fuel	• Power stations provide jobs. • Provides energy for homes and industry. • Does not take up much space.	• Produces pollutants. • Causes global warming due to the greenhouse effect. • Non-renewable source so is in danger of running out.
Ethanol	• Does not affect the performance of a car. • Can save money. • Can be made from renewable resources. • Less carbon emissions. • Less carbon monoxide produced.	• Need to pay out to convert engines. • Much more sugar will need to be grown to meet demand. • Price of sugar is likely to rise due to increased demand. • Produces less energy than petrol when it burns.
Hydrogen	• No harmful gases produced, only water vapour which does not harm the environment.	• Expensive to produce (requires lots of energy). • Difficult to store and transport safely.

C1.5 Other useful substances from crude oil

Fractions produced from the distillation of crude oil can be cracked to produce alkenes such as ethene. Alkenes can be used to make polymers and ethanol. To understand this, you need to know:

- how hydrocarbons are cracked
- how fuels and the starting materials for polymers can be produced from cracking
- how polymers are formed and their properties and uses
- how ethanol is produced from both renewable and non-renewable resources.

Cracking Hydrocarbons

Long-chain hydrocarbons can be cracked (broken down) into shorter chains using heat. When long-chain hydrocarbons are cracked, a short-chain alkane and alkene are produced. The short-chain alkane is more useful as a fuel and the short-chain alkene can be used to make polymers.

Hydrocarbons are cracked by heating them until they vaporise, then passing the vapour over a heated catalyst, where a thermal decomposition reaction takes place.

Apparatus Used for Cracking in the Laboratory

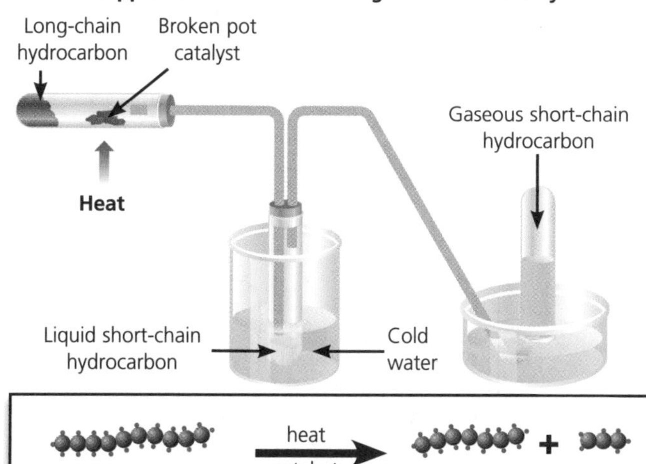

Hydrocarbons can also be cracked by mixing them with steam and heating them to very high temperatures.

Alkenes (Unsaturated Hydrocarbons)

We have already seen that carbon atoms can form single bonds with other atoms, but they can also form double bonds. Some of the products of cracking are hydrocarbon molecules with at least one double bond; this is an **unsaturated** hydrocarbon and it is known as an alkene. Alkenes always have names ending in 'ene'.

The general formula for alkenes is C_nH_{2n}. The simplest alkene is ethene, C_2H_4, which is made up of 4 hydrogen atoms and 2 carbon atoms. As you can see in the diagram below, ethene contains one double carbon–carbon bond (often represented by '=').

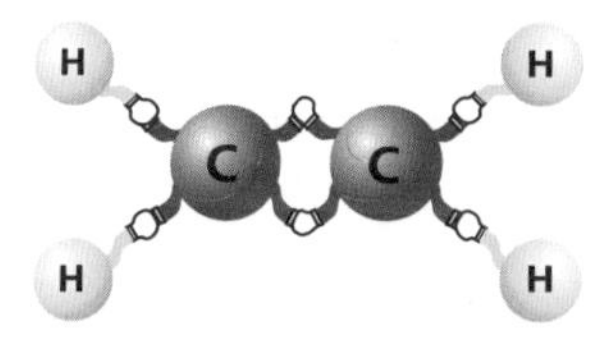

Here is another way of representing alkenes:

Name	Formula	Displayed Formula
Ethene	C_2H_4	H H C=C H H
Propene	C_3H_6	H H C=C–C–H H H H

Not all the carbon atoms are linked to 4 other atoms; a double carbon–carbon bond is present instead. Because they are unsaturated, they are useful for making other molecules, especially **polymers**.

When alkenes react with orange bromine water it turns colourless. This reaction can be used to distinguish between alkanes and alkenes because alkanes do not react with bromine water in this way.

Polymerisation

Because alkenes are unsaturated (have a double bond), they are very reactive. When small alkene molecules (monomers) join together to form long-chain molecules (polymers), it is called **polymerisation**.

The properties of polymers depend on what they are made from and the conditions under which they are made. The materials commonly called plastics are all synthetic polymers. They are produced commercially on a very large scale and have a wide range of properties and uses. Polymers and plastics were first discovered in the early 20th century.

Making Poly(ethene) from Ethene

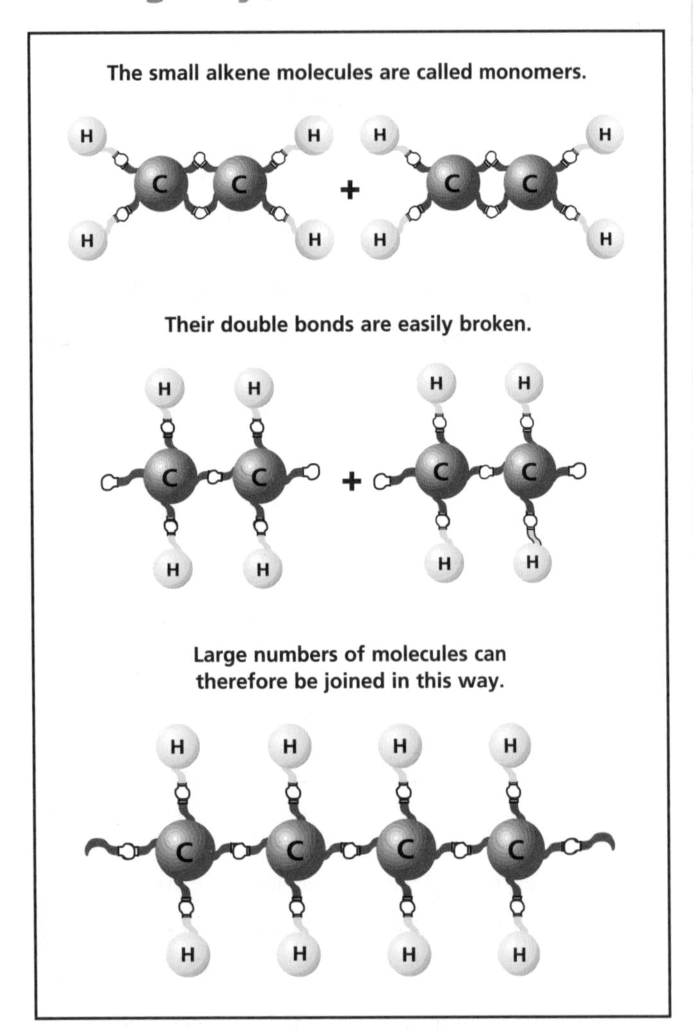

The resulting long-chain molecule is a polymer – in this case poly(ethene), often called polythene. Poly(propene) can be made in a similar way.

Representing Polymerisation

A more convenient way of representing polymerisation is:

Ethene monomers (unsaturated) → **Poly(ethene) polymers (saturated)**

$$\begin{array}{c} H\quad H \\ |\quad\;\; | \\ C=C \\ |\quad\;\; | \\ H\quad H \end{array} + \begin{array}{c} H\quad H \\ |\quad\;\; | \\ C=C \\ |\quad\;\; | \\ H\quad H \end{array} + \longrightarrow \begin{array}{c} \;\;H\quad\; H\quad\; H\quad\; H \\ \;\;|\quad\;\;\; |\quad\;\;\; |\quad\;\;\; | \\ -C-C-C-C- \\ \;\;|\quad\;\;\; |\quad\;\;\; |\quad\;\;\; | \\ \;\;H\quad\; H\quad\; H\quad\; H \end{array}$$

...and thousands more... **...and on and on...**

General Equation for Polymerisation

This equation can be used to represent the formation of any simple polymer:

$$n\left(\begin{array}{c} |\quad\;\; | \\ C=C \\ |\quad\;\; | \end{array}\right) \longrightarrow \left(\begin{array}{c} |\quad\;\; | \\ C-C \\ |\quad\;\; | \end{array}\right)_n$$

where n is a very large number

For example, if we take n molecules of propene we can produce poly(propene), which is used to make crates and ropes:

$$n\left(\begin{array}{c} H\quad CH_3 \\ |\quad\;\; | \\ C=C \\ |\quad\;\; | \\ H\quad H \end{array}\right) \xrightarrow[\text{catalyst}]{\text{pressure}} \left(\begin{array}{c} H\quad CH_3 \\ |\quad\;\; | \\ C-C \\ |\quad\;\; | \\ H\quad H \end{array}\right)_n$$

And n molecules of chloroethene can produce polychloroethene (also known as polyvinyl chloride or PVC):

$$n\left(\begin{array}{c} H\quad Cl \\ |\quad\;\; | \\ C=C \\ |\quad\;\; | \\ H\quad H \end{array}\right) \xrightarrow[\text{catalyst}]{\text{pressure}} \left(\begin{array}{c} H\quad Cl \\ |\quad\;\; | \\ C-C \\ |\quad\;\; | \\ H\quad H \end{array}\right)_n$$

Polymers are classified by the reactions by which they are formed. These are examples of **addition** polymers, formed by addition polymerisation.

Uses of Polymers

Specific polymers can have different uses, for example:

- **Polyvinyl chloride (PVC)** can be used to make waterproof items and drain pipes and can also be used as an electrical insulator.
- **Polystyrene** is used to make casings for electrical appliances and it can be expanded to make protective packaging.
- **Poly(ethene)** is commonly used to make plastic bags and bottles.
- **Poly(propene)** can be used to make crates and ropes.

Polymers have many useful applications and new uses are being developed.

Polymers and composites are widely used in medicine and dentistry:

- Implantable materials are used for hard and soft tissue surgery, replacing and fusing damaged bone and cartilage.
- Hard-wearing anti-bacterial dental cements, coatings and fillers have been produced.
- Hydrogels can be used as wound dressings.
- Silicone hydrogel contact lenses have been developed over the last few years. Research has shown that people who wear this type of contact lens have a 5% lower risk of developing severe eye infections.

Polymers and composites can be used to coat fabrics with a waterproof layer. Smart materials, including shape-memory polymers, are also increasingly more common.

Disposing of Plastics

Because plastic is such a versatile material and it is cheap and easy to produce, we tend to generate a large amount of plastic waste.

The main problem with most plastics is that they are not biodegradable. This means that if they are left as litter or buried in landfill sites they cannot be broken down by microbes and will therefore not decompose and rot away. This leads to a build-up of plastic waste in landfill sites.

Plastics can also be burned but this produces air pollution, including toxic fumes from some plastics.

One way in which scientists are trying to overcome the problem with the disposal of plastics is to make biodegradable polymers. An example of this is a polymer made from cornstarch.

Plastics are recycled in order to reduce the amount sent to landfill.

Making Alcohol from Ethene

Ethanol is an **alcohol**. It can be produced by reacting steam with ethene at a moderately high temperature and pressure in the presence of a catalyst, phosphoric acid.

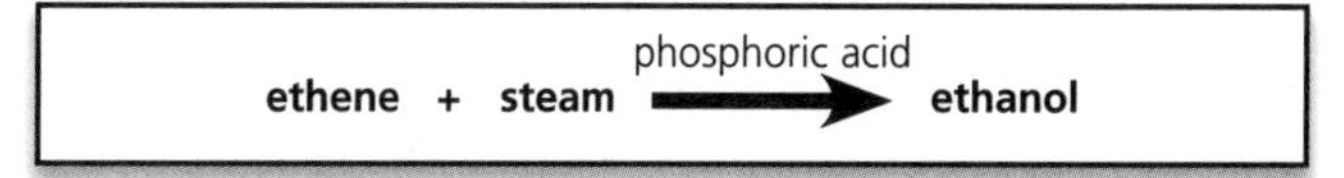

Ethanol can be used as:

- a solvent
- a fuel.

Producing ethanol from ethene means using a non-renewable resource (crude oil) and this is one of the main disadvantages of this method.

Making Alcohol from Sugar

Ethanol can also be produced from sugar, which is a renewable resource. This process is called fermentation and is carried out in solution using yeast at temperatures of about 37°C.

The process produces ethanol and carbon dioxide.

Fermentation is used to make alcoholic drinks but it can also be used to make ethanol to burn as a fuel.

You need to be able to evaluate the social and economic advantages and disadvantages of using products from crude oil as fuels, or as raw materials for plastic and other chemicals.

Crude oil is one of our most important natural resources. It is hard to imagine what our lives would be like without the products that we can get from crude oil. Transport would come to a standstill, there would be no more plastics and detergents, and the pharmaceutical industry would not be able to get essential raw materials, so medicines would run out.

Crude oil can be used to make tough, lightweight, waterproof and breathable fabrics for clothes; paint for cars; dyes; packaging and communication equipment. However, it is important to weigh up the advantages of the products we can get from crude oil against the disadvantages of using it as a raw material, particularly since it is a limited resource.

Advantages of Crude Oil
• The oil industry provides jobs. • The fractions of crude oil have many uses. • Provides raw materials for industry. • Provides fuel for transport.
Disadvantages of Crude Oil
• Oil spills damage the environment. • Can cause air pollution. • Increases global warming. • Produces non-biodegradable material.

You need to be able to consider and evaluate the social, economic and environmental impacts of the uses, disposal and recycling of polymers.

Plastics (polymers) are everywhere. There are a wide range of polymers with different, highly useful physical properties – some polymers are flexible, others are rigid; some have a low density, whereas others are very dense. They can be transparent or opaque. They are waterproof and resistant to corrosion and can be used as a protective layer.

However, although polymers are relatively cheap to produce, the cost to society and the environment has to be considered. Pollution, and its effects on residents who live near polymer-producing factories, is a major issue. And the problem of disposal of polymers once they have been used needs to be addressed because burning them produces harmful and sometimes toxic gases.

Advantages of Polymers
• Cheap to make. • Many uses because of their different properties. • Provide jobs in factories that make the polymers and the products. • Some polymers can be recycled, melted down and made into something else, which saves valuable natural resources. • If polymers are used instead of wood, fewer trees will have to be cut down.
Disadvantages of Polymers
• People do not like to live near polymer-producing industrial works. • Some people think plastic products look cheap compared with natural materials. • Made from oil, a non-renewable resource. • Most plastics are not biodegradable so there is a problem of how to get rid of them. • Give off toxic fumes when they burn. • Sorting types of polymers for recycling can be expensive.

You need to be able to evaluate the advantages and disadvantages of making ethanol from renewable and non-renewable sources.

Using Non-renewable Sources

Ethanol can be produced by reacting steam with ethene at a moderately high temperature and pressure in the presence of the catalyst, phosphoric acid.

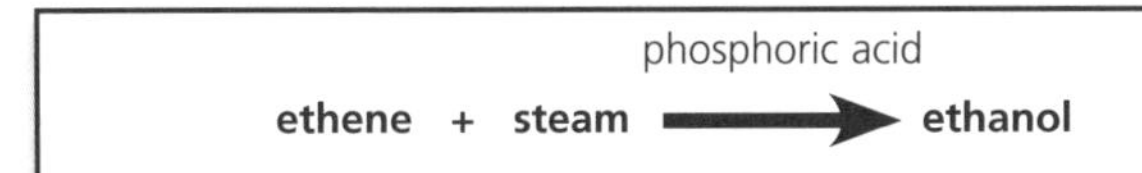

Using Renewable Sources

Ethanol can also be produced by the fermentation of sugars. Water and yeast are mixed with the raw materials at just above room temperature. Enzymes, which are biological catalysts found in the yeast, react with the sugars to form ethanol and carbon dioxide. The carbon dioxide is allowed to escape from the reaction vessel, but air is prevented from entering it. The ethanol is separated from the reaction mixture by fractional distillation when the reaction is over.

One problem with the production of ethanol is that it can be oxidised by air (in certain conditions) to produce ethanoic acid. The presence of ethanoic acid results in alcoholic drinks turning sour.

Method	Advantages	Disadvantages
Reacting ethene with steam	• Fast rate of production. • High-quality ethanol produced. • Can be produced continuously. • Best method for making large quantities.	• Uses non-renewable raw material (crude oil). • High temperatures needed.
Fermentation	• Renewable raw material (sugar). • Fairly high-quality ethanol produced after fractional distillation. • Best method for making small quantities. • Lower temperatures needed.	• Slow rate of production. • Is produced in batches. • Ethanol not as pure as that produced by reacting ethene with steam.

Plant Oils and their Uses

C1.6 Plant oils and their uses

Many plants produce oils that can be extracted and used for many purposes. Vegetable oils can be hardened to make margarine or to make biodiesel fuel. Emulsions can be made and have a number of uses. To understand this, you need to know:

- how oils are extracted from plants
- the properties and uses of oils
- what emulsions are.

Getting Oil from Plants

Many plants produce fruit, seeds and nuts that are rich in **oils**, which can be extracted and made into consumer products. Some common examples you might find in the food you eat are:

- sunflower oil
- olive oil
- oilseed rape
- palm kernel oil.

Oil can be extracted from plant materials by pressing (crushing) them or by distillation. This removes the water and other impurities from the plant material.

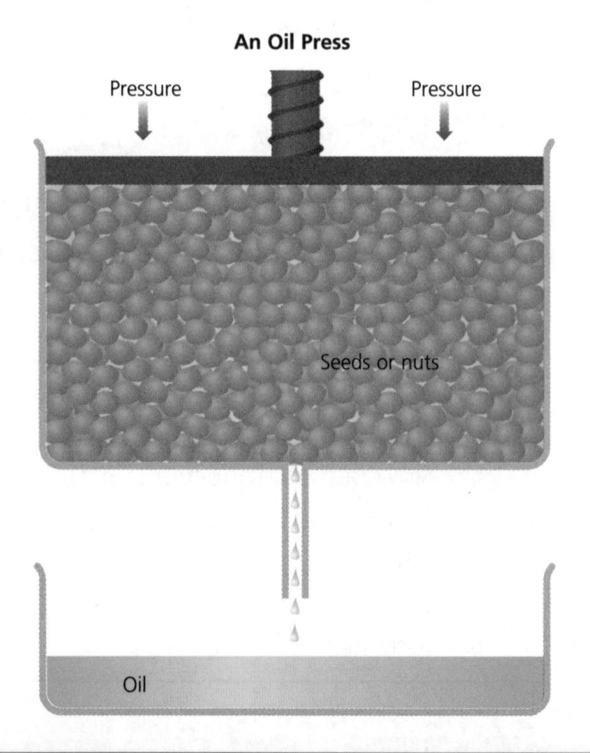

Vegetable Oils

Vegetable oils are important **foods** and **fuels** because they provide nutrients and a lot of energy. Some vehicles can now be converted to use vegetable oils as their fuel instead of petrol or diesel.

Vegetable oils contain double carbon–carbon bonds, so they are described as **unsaturated**. They can be detected using bromine water. They react with the orange-coloured bromine water to decolourise it. The bromine becomes part of the compound, by breaking the double bond. For example:

$$H_2C{=}CH_2 + Br_2 \longrightarrow H{-}CHBr{-}CHBr{-}H$$

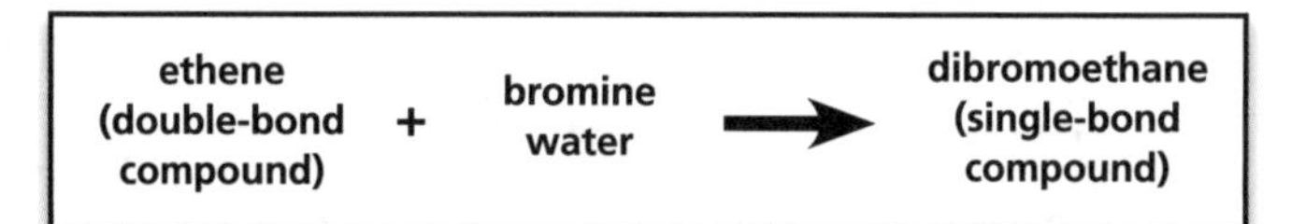

ethene (double-bond compound)	+	bromine water	→	dibromoethane (single-bond compound)

They will react in the same way with iodine (I_2).

Vegetable oils are used to cook foods because they boil at higher temperatures than water and, therefore, allow food to be cooked at higher temperatures. This changes both the texture and taste of food. Cooking at higher temperatures in oil means that food is cooked more quickly. It also increases the energy that the food produces when eaten because it is coated in oil.

Emulsions

Oils do not dissolve in water. Oil and water have different densities. A mixture of oil and water is called an **emulsion**. If oil and water are mixed thoroughly, droplets of oil can be seen dispersed in the water.

Emulsions are thicker than oil or water and have a better texture, appearance and coating ability. They have many uses, e.g. salad dressing and ice cream.

If you mix some olive oil and vinegar together you can make a salad dressing. However, it does not stay mixed for very long because the water particles in the vinegar clump together and the oil particles clump together. The mixture can be seen to separate into two layers.

You can make a salad dressing last longer by adding some mustard to it before shaking the mixture up. This stops the separate layers forming. The mustard is an **emulsifying agent**.

Many different emulsifying agents are used in the manufacture of food to stop vegetable oils and water forming separate layers.

HT Emulsifiers are used to stop oil and water separating, for example egg yolks are used as emulsifiers in mayonnaise.

Emulsifiers work because different parts of the molecule are attracted to the two different liquids. The 'heads' of the emulsifiers are attracted to the water and the 'tails' are attracted to the oil.

Simple Model of an Emulsifier

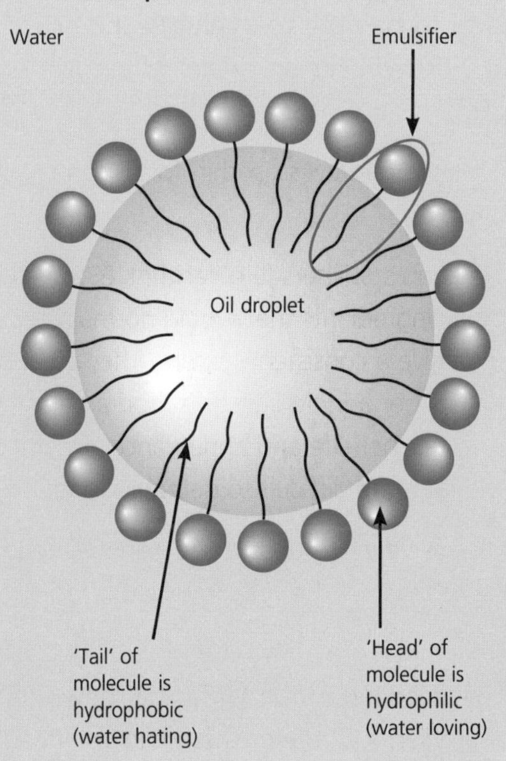

HT The Manufacture of Margarine

As a general guide, the more double carbon–carbon bonds present in a substance, the lower its melting point. This means that unsaturated fats, e.g. vegetable oils, tend to have melting points below room temperature and are called oils.

For some purposes you might need a solid fat, for example, to spread on your bread or to use to make cakes and pastries. You can raise the melting point of an oil to above room temperature by removing some or all of the double carbon–carbon bonds.

When ethene and hydrogen are heated together in the presence of a nickel catalyst, a reaction takes place that removes the double carbon–carbon bonds to produce ethane. This process is called **hydrogenation** and is used to convert unsaturated hydrocarbons into more saturated hydrocarbons.

$$H_2C{=}CH_2 + H_2 \xrightarrow[\text{catalyst}]{\text{nickel}} H_3C{-}CH_3$$

unsaturated fat + **hydrogen** $\xrightarrow[\text{catalyst}]{\text{nickel}}$ **saturated fat**

Margarine is manufactured from unsaturated vegetable oils like sunflower oil. The oil is reacted with hydrogen, at a temperature of around 60°C in the presence of a nickel catalyst and some of the double bonds are hydrogenated. Removing more double bonds makes the margarine harder. Unfortunately, hydrogenation of vegetable oils also makes them less healthy to eat.

Margarine

You need to be able to evaluate the effects of using vegetable oils in food and the impacts on diet and health.

Oils have many uses. However, the amount of saturated fats we consume needs to be carefully controlled to reduce the risk of heart disease.

Vegetable oils are a healthy alternative to using fats derived from animals because they contain monounsaturated fats that can lower blood cholesterol levels, and they contain no cholesterol.

However, it is important to remember that you should not consume too much of any oil because this would not lead to a healthy balanced diet. Where possible, unsaturated fats such as olive oil should be used to reduce the health risks. However, using saturated fats occasionally would not be too bad for your health.

You need to be able to evaluate the benefits, drawbacks and risks of using vegetable oil to produce fuels.

Cars can now be converted so that they can run on vegetable oil instead of diesel. Although in many ways this is a more environmentally friendly option, it has not yet become widespread, and only a few cars have been converted.

Advantages of Vegetable Oils as Fuels
• Cheaper than diesel or petrol. • Fewer pollutant gases produced – virtually carbon neutral. • No change to performance of car. • Renewable source.
Disadvantages of Vegetable Oils as Fuels
• High cost of conversion kit. • Unpleasant smell. • Inconvenience of filling up your car – vegetable oil currently not an option at petrol stations. • Increased demand may put up prices for food made using vegetable oil. • Large areas of land needed to grow oil crops.

You need to be able to evaluate the use, benefits, drawbacks and risks of emulsifiers in foods.

Emulsifiers are additives used in foods to produce emulsions.

Emulsifiers and stabilisers (e.g. lecithin, E322) are used to mix ingredients that would normally separate, in order to give a consistent texture. They are used along with other additives to give food a better texture, taste, shelf-life and appearance. Emulsifiers make the fat more difficult to detect in foods.

Additives are very useful, but the benefits need to be weighed against the risks involved in using them in food.

Any additive added to food must be shown on the list of ingredients. Some of the additives added to food have been given E-numbers.

C1.7 Changes in the Earth and its atmosphere

The Earth is made up of four layers and is surrounded by an atmosphere that both protects us and provides the conditions needed for life. The atmosphere has changed drastically since the Earth was first formed, but has remained constant for the last 200 million years. Recent human activities have caused further changes in the atmosphere. To understand this, you need to know:

- the structure of the Earth
- what tectonic plates are, how they move and what this causes
- what the atmosphere consists of and how it was formed from the early atmosphere.

Structure of the Earth

The Earth is nearly spherical and has a layered structure that consists of:

- a thin crust – thickness varies between 10km and 100km
- a mantle – extends almost halfway to the centre and has all the properties of a solid even though it does flow very slowly
- an inner and outer core (made of nickel and iron) – over half of the Earth's radius. The outer core is liquid and the inner core solid.

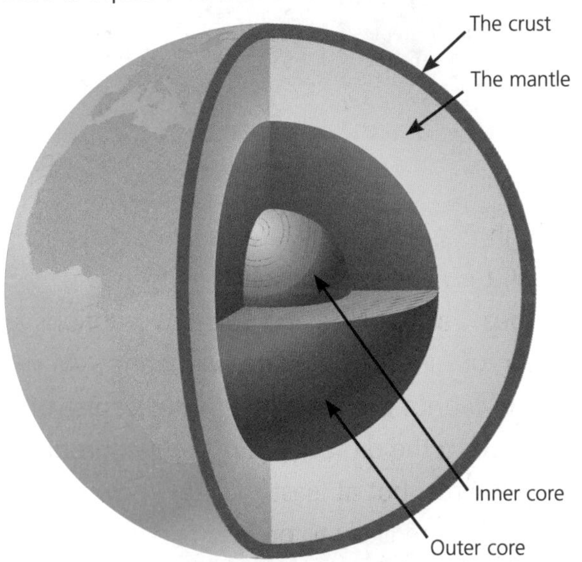

The average density of the Earth is much greater than the average density of the rocks that form the crust, because the interior is made of a different, denser material than that of the crust.

Although there does not seem to be much going on, the Earth and its crust are very dynamic. Rocks at the Earth's surface are continually being broken up, reformed and changed in an ongoing cycle of events known as the rock cycle. It is just that the changes take place over a very long time.

Tectonic Theory

At one time people used to believe that features on the Earth's surface were caused by shrinkage as the Earth cooled, following its formation. However, as scientists have found out more about the Earth, this theory has now been rejected.

A long time ago, scientists noticed that the east coast of South America and the west coast of Africa have:

- similar patterns of rocks, which contain fossils of the same plants and animals, e.g. the Mesosaurus

- closely matching coastlines.

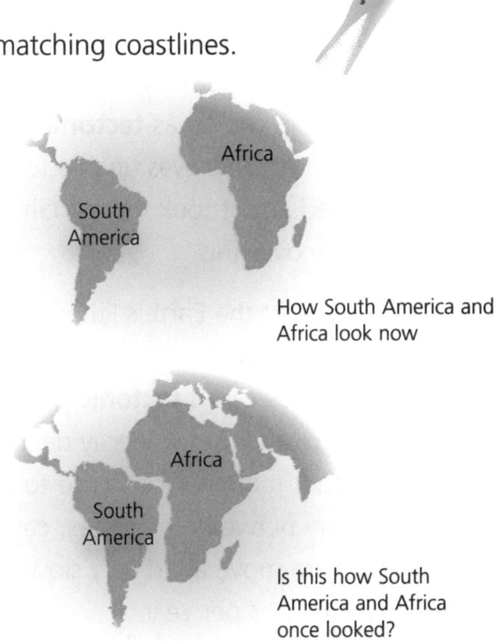

Tectonic Theory (cont)

This evidence led Alfred Wegener to propose that, even though they are now separated by thousands of kilometres of ocean, South America and Africa had at one time been part of a single land mass.

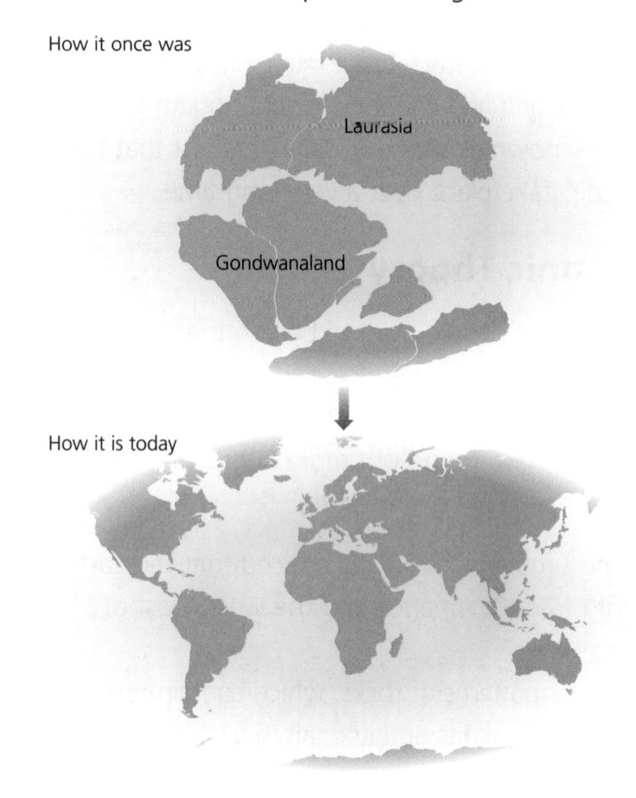

He proposed that the movement of the crust was responsible for the separation of the land (continental drift), which explains the movement of the continents from where they were (as Gondwanaland and Laurasia) to how they look today. This is known as **tectonic theory**. Unfortunately, Wegener was unable to explain *how* the crust moved and it took more than 50 years for scientists to discover this.

We now know that the Earth's lithosphere (the crust and the upper part of the mantle) is 'cracked' into several large pieces called **tectonic plates**. Intense heat, released by radioactive decay deep in the Earth, causes hot molten rock to rise to the surface at the boundary between the plates, causing the tectonic plates to move apart very slowly, at speeds of a few centimetres per year.

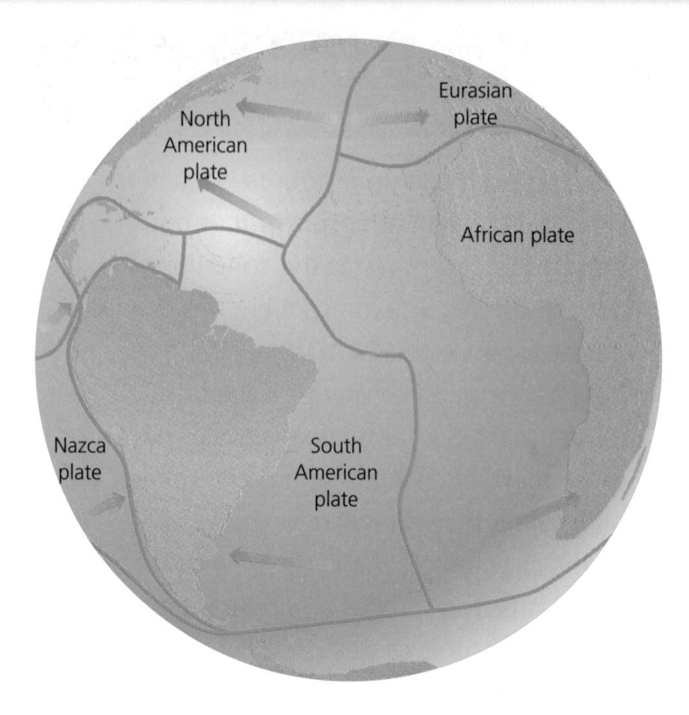

In convection in a gas or a liquid, the matter rises as it is heated, then as it gets further away from the heat source it cools and sinks down again. The same happens in the mantle. The hot molten rock rises to the surface, creating new crust. The older crust, which is cooler, then sinks down where the convection current starts to fall. This causes the land masses on these plates to move slowly across the globe.

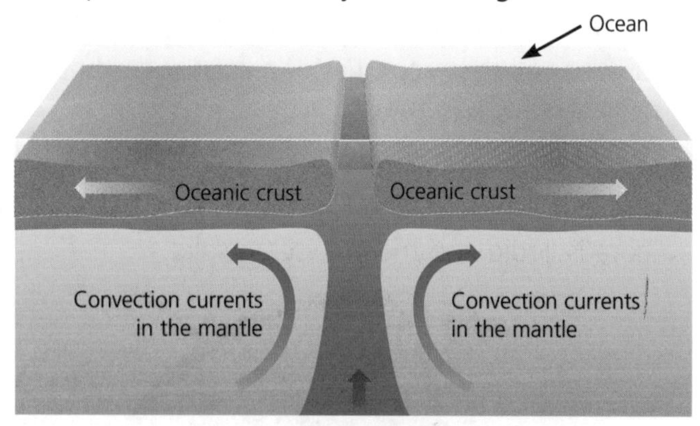

Although the movements are usually small and gradual, they can sometimes be sudden and disastrous. Earthquakes and volcanic eruptions are common occurrences at plate boundaries. As yet, scientists cannot predict *when* these events will occur, due to the difficulty in making appropriate measurements, but at least they do know *where* these events are likely to occur.

The Earth's Atmosphere

Since the formation of the Earth 4.6 billion years ago the atmosphere has changed a lot. The timescale, however, is enormous because one billion years is one thousand million (1 000 000 000) years!

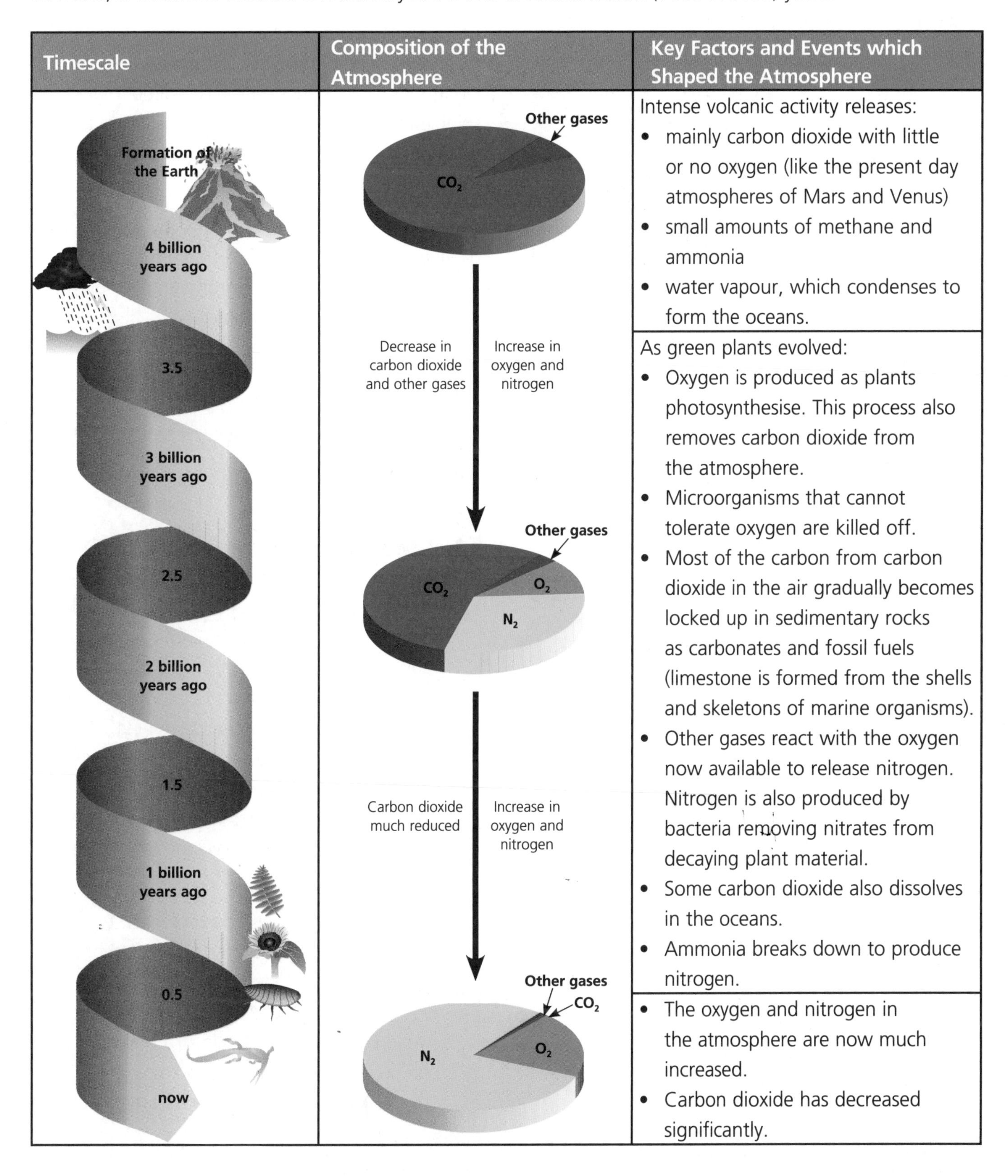

Timescale	Composition of the Atmosphere	Key Factors and Events which Shaped the Atmosphere
		Intense volcanic activity releases: • mainly carbon dioxide with little or no oxygen (like the present day atmospheres of Mars and Venus) • small amounts of methane and ammonia • water vapour, which condenses to form the oceans.
		As green plants evolved: • Oxygen is produced as plants photosynthesise. This process also removes carbon dioxide from the atmosphere. • Microorganisms that cannot tolerate oxygen are killed off. • Most of the carbon from carbon dioxide in the air gradually becomes locked up in sedimentary rocks as carbonates and fossil fuels (limestone is formed from the shells and skeletons of marine organisms). • Other gases react with the oxygen now available to release nitrogen. Nitrogen is also produced by bacteria removing nitrates from decaying plant material. • Some carbon dioxide also dissolves in the oceans. • Ammonia breaks down to produce nitrogen.
		• The oxygen and nitrogen in the atmosphere are now much increased. • Carbon dioxide has decreased significantly.

Composition of the Atmosphere

Our atmosphere has been more or less the same for about 200 million years. The pie chart below shows how it is made up. Water vapour may also be present in varying quantities (0–3%).

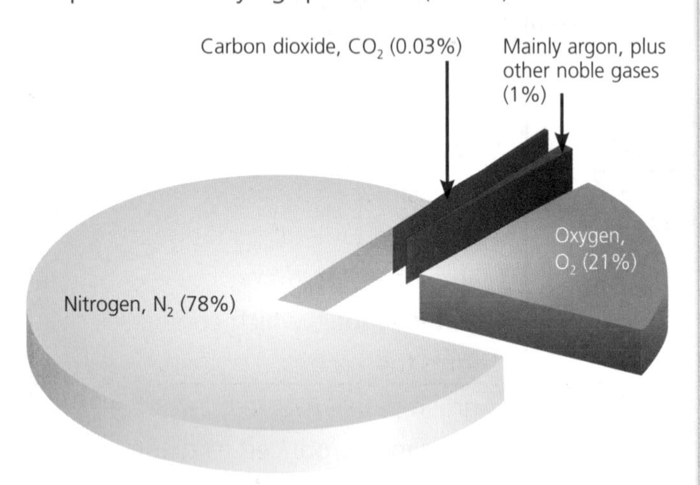

The noble gases (in Group 0 of the Periodic Table) are all chemically unreactive gases and are used in light bulbs and electric discharge tubes. Helium is much less dense than air and is used in balloons.

HT Air is a mixture containing different gases. These gases are used as a source of raw materials in a variety of industrial processes. To separate the gases we use fractional distillation because they all have different boiling points.

Changes to the Atmosphere

The level of carbon dioxide in the atmosphere today is increasing due to:

- volcanic activity – geological activity moves carbonate rocks deep into the Earth. During volcanic activity they may release carbon dioxide back into the atmosphere.
- burning of fossil fuels – burning carbon, which has been locked up in fossil fuels for millions of years, releases carbon dioxide into the atmosphere.

The increase in levels of carbon dioxide in the atmosphere, particularly from the burning of fossil fuels, is thought to cause global warming.

The level of carbon dioxide in the atmosphere is reduced by the reaction between carbon dioxide and sea water. Increased carbon dioxide in the atmosphere increases the rate of the reaction between carbon dioxide and sea water. This reaction produces insoluble carbonates (mainly calcium), which are deposited as sediment, and soluble hydrogencarbonates (mainly calcium and magnesium). The carbonates form sedimentary rocks in the Earth's crust.

Although the oceans act as a reservoir for carbon dioxide, increased levels of carbon dioxide do have an impact on the marine environment. Dissolving CO_2 lowers the pH of seawater.

HT Origin of Life

There are many different theories of how life on Earth began. At the moment there is no single accepted theory of how life originated on Earth.

One possible theory is known as the **Primordial Soup Theory**. This theory suggests that life on Earth began in an ocean or pond when a form of energy (lightning) caused the combination of chemicals from the atmosphere (hydrocarbons, ammonia and water) to combine and make amino acids. Amino acids are the building blocks of proteins. The primordial soup was an environment where, when the first living things were formed, they had all they needed to survive and reproduce.

In the 1950s two scientists, Stanley Miller and Harold Urey, carried out an experiment to test this theory. By using electricity to represent the lightning, they combined the gases thought to be present in the Earth's early atmosphere. They did generate amino acids from this experiment – however, there are many problems with the theory because amino acids are not living material.

You need to be able to explain why the theory of crustal movement (continental drift) was not generally accepted for many years after it was proposed.

About 200 years ago, most geologists thought that the Earth had gone through a period of being extremely hot and consequently had dried out and contracted, or shrunk, as it cooled. Features of the Earth, such as mountain ranges, were thought to have been wrinkles that formed in the Earth's crust as it shrank.

At this point, insufficient data had been collected to show that the continents were in fact moving, and nobody produced any evidence to contradict the theory of the Earth shrinking until the early 1900s.

Then Alfred Wegener studied certain features of the Earth (see p.71–72), which prompted him to propose his theory of continental drift in 1915. This theory proposed that the Earth is made up of plates that have moved slowly apart. Most geologists at the time said that this theory was impossible, although a few did support Wegener.

In the 1950s scientists were able to investigate the ocean floor and found new evidence to support Wegener's theory. They discovered that although he was wrong about some aspects, the basis for his theory was correct.

By the 1960s, geologists were convinced by the theory of continental drift and can now use it to explain many geological features and occurrences caused by moving tectonic plates. Evidence now shows that the sea floor is spreading outwards and convection currents in the mantle cause movement of the crust.

You need to be able to explain why scientists cannot accurately predict when earthquakes and volcanic eruptions will occur.

To understand how earthquakes and volcanic eruptions occur, we need to consider the movement of the tectonic plates. They can stay in the same position for some time, resisting a build up of pressure, and then when the pressure becomes too great they can suddenly move.

However, it is impossible to predict exactly when this will happen because the plates do not move in regular patterns. Scientists can measure the strain in underground rocks to see if they can calculate when an earthquake is likely to happen, but they are unlikely to be able to give an exact forecast.

A volcano erupts when molten rock rises up into the spaces between the rocks near the surface. Scientists have instruments which can identify these changes, and therefore warn of imminent eruptions. However, sometimes the molten rock cools, so the magma does not reach the surface and the volcano does not erupt. So, other factors which are hard to predict can affect whether a volcano erupts or not.

Therefore, despite having very sophisticated equipment which monitors volcanic activity and areas prone to earthquakes, scientists cannot always predict exactly when they might happen.

You need to be able to explain and evaluate the effects of human activities on the atmosphere.

Most human activities, especially those used to create heat and energy, can produce pollutants that are harmful to the atmosphere. Some of these activities and their impact on the atmosphere are listed below.

The burning of fossil fuels create the gases sulfur dioxide and carbon dioxide. Sulfur dioxide contributes to the formation of acid rain, which can erode buildings and add acid to lakes and the soil.

The carbon dioxide content of the air used to be roughly constant (0.03%) but it has been increased by the growth in population, which raises energy requirements.

Deforestation means less photosynthesis takes place. An increase in the level of carbon dioxide is believed to be responsible for global warming and climate change.

The combustion of petrol and diesel allows the reaction of nitrogen and oxygen at the very high temperatures in car engines to make oxides of nitrogen, which are pollutants. Carbon monoxide is produced by the incomplete combustion of fuels. It is a poisonous gas which is eventually oxidised to carbon dioxide, adding to global warming.

So, what can we do to reduce the pollutants? We need to create energy to heat homes, power our cars, etc., but we also need to look at the effect that various fuels have on the environment and consider alternative methods of producing energy in order to limit the impact we have on the planet.

- Cars fitted with catalytic converters reduce the production of carbon monoxide and oxides of nitrogen.
- Alternative forms of energy can reduce pollution, e.g. wind farms and hydroelectricity.
- Power stations can use fuels that reduce atmospheric pollution, e.g. nuclear power stations.

The European Union and the United Kingdom have made laws to control the pollution levels. These levels need to be regulated and monitored, and every country needs to control their pollution levels.

Exam Practice Questio

1 **a)** Draw the electronic structure for a chlorine atom. **(1**

b) Explain why chlorine is in Group 7 of the Periodic Table. **(1 mar**

2 **a)** What is meant by the term 'ore'? **(1 mark)**

b) Metals are very useful in everyday life. Use the words in the box to complete the sentences below (you may not need to use all the words). **(3 marks)**

electrolysis	**less**	**more**	**oxidation**	**reduction**

Metals can be extracted from their ores using a process called .... This involves removing the oxygen from the metal oxide by heating it with another element. Metals that are .... reactive than carbon can be extracted from their ores by being heated with carbon. Metals that are more reactive than carbon are extracted from their ores by a process called ....

3 Crude oil is a mixture of different hydrocarbons. It is a very useful raw material. Most of the hydrocarbons in crude oil are alkanes.

a) Which elements are found in a hydrocarbon? **(1 mark)**

b) What is the general formula for an alkane? **(1 mark)**

c) Long-chain alkanes are less useful than short-chain alkanes. Long-chain alkanes can be broken down to make more useful materials.

i) What are the two conditions required for this reaction to occur? **(2 marks)**

ii) Give the name of this reaction. **(1 mark)**

iii) What type of compound is produced in this reaction along with a shorter-chain alkane? **(1 mark)**

iv) Explain a chemical test to distinguish between the short-chain alkane and the other product of cracking (**part iii**). In your answer you should give the name of any chemical used and what observation you would make. **(2 marks)**

HT 4 Balance these symbol equations:

a) $Na + O_2 \longrightarrow Na_2O$ **(1 mark)**

b) $H_2SO_4 + NaOH \longrightarrow Na_2SO_4 + H_2O$ **(1 mark)**

...nergy by heating ...e factors that ...: which that ...rred

Thermal energy can be transferred by conduction, convection, infrared radiation, evaporation and condensation. You need to know which processes are important in particular situations and be able to explain how to reduce heat transfer in vacuum flasks, buildings, humans and animals. To understand this, you need to know:

- the role played by particles in heat transfer
- the main ways of insulating a home
- why silvered surfaces are used in vacuum flasks
- why trapped air is a good insulator.

Infrared Radiation

Thermal (infrared) **radiation** is the **transfer** of **thermal energy** (heat energy) by electromagnetic waves; no particles of matter are involved.

All objects emit and absorb thermal radiation. The hotter the object, the more energy it radiates. The amount of thermal radiation given out or taken in by an object depends on its surface, shape and dimensions.

An object will emit or absorb energy faster if there is a big difference in temperature between it and its surroundings. The rate of heat transfer can be slowed down by the use of insulation, which provides a barrier.

Under similar conditions, different materials transfer heat at different rates. At the same temperature:

- dark matt surfaces emit more radiation than light shiny surfaces
- dark matt surfaces absorb more radiation than light shiny surfaces because shiny surfaces are good reflectors of infrared radiation.

Kinetic Theory

The particles in solids, liquids and gases have different amounts of energy.

In a solid the particles have the least energy. They are not able to move but do vibrate around a fixed point.

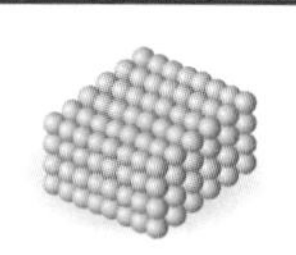

Heating a solid gives the particles enough energy to move around (although they cannot move far apart) and it melts to become a liquid.

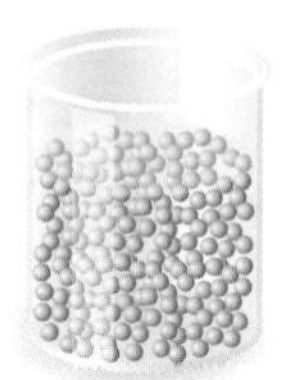

Further heating gives the particles enough energy to move very quickly. They separate from each other and the liquid evaporates to become a gas.

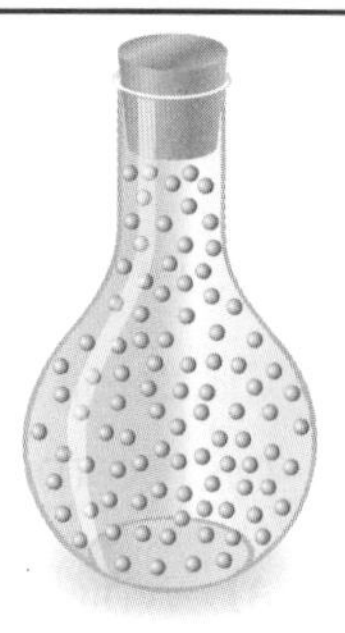

Conduction

Conduction is the transfer of heat energy without the substance itself moving.

The structure of metals makes them good conductors of heat. As a metal becomes hotter, its tightly packed particles gain more kinetic energy and vibrate. This energy is transferred to cooler parts of the metal by delocalised electrons, which move freely through the metal, colliding with particles and other electrons.

N.B. An electron is a subatomic particle.

Conduction also occurs in non-metal solids because the particles can pass energy from one to the next by vibration. However, the lack of free electrons makes most non-metals poor conductors. Gases are bad conductors because the particles are so far apart.

Heat energy is conducted up the poker as the hotter parts transfer energy to the colder parts

Convection

Convection is the transfer of heat energy through movement. This occurs in liquids and gases and creates convection currents.

In a liquid or gas the particles nearest the heat source move faster and become further apart. This causes the substance to expand and become less dense than the colder parts.

The warm liquid or gas rises up and colder, denser liquid or gas moves into the space created (close to the heat source).

Example 1

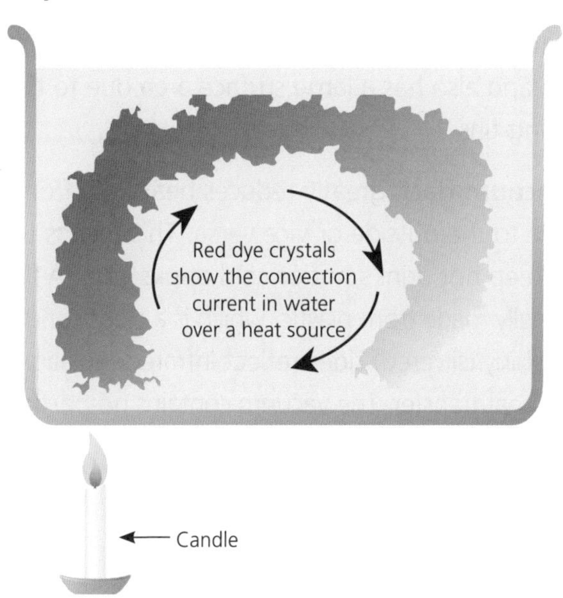

Example 2

Circulation of air caused by a radiator

Air cools, becomes denser and sinks

Air warms up, becomes less dense and rises

Cooler air replaces air which has risen

Condensation

The particles in a gas have more energy than those in a liquid. When a gas condenses to form a liquid, this energy is released and can make the temperature increase.

Condensation can occur when a warm gas comes in contact with a cold surface, e.g. water vapour in breath condensing on a window.

The surface needs to be cold enough to cool the particles so that they no longer have enough energy to move around quickly as in a gas. The colder the surface the greater the rate of condensation.

Evaporation

A liquid evaporates when its particles have gained enough energy to escape the surface of the liquid and become a gas.

For example, you feel cold when you get out of a swimming pool because the water is evaporating from your body and in doing so it takes heat energy from you.

Clothes on a washing line dry quickest on a sunny and windy day because the Sun provides the energy for the particles to escape from the surface, and the wind blows them away (making space for more water particles to escape).

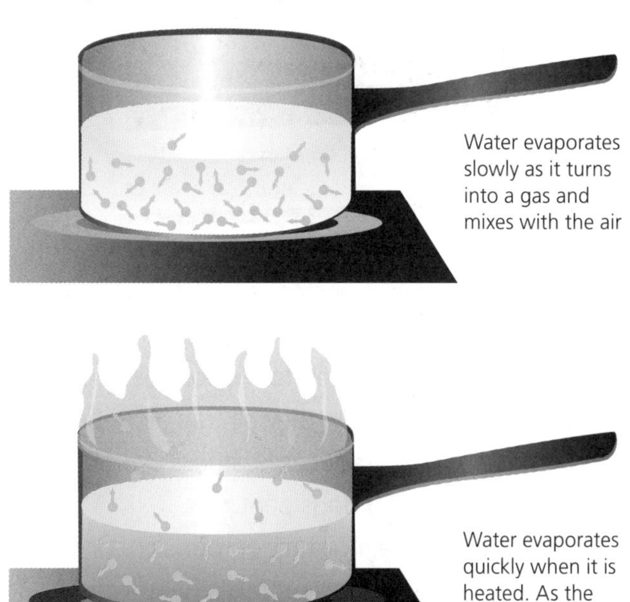

You need to be able to apply ideas about energy transfer to keeping warm or cool.

The rate at which heat is transferred depends on a number of factors:

1. A large surface area compared to volume will gain and lose heat quicker.
   Example – desert foxes have big, thin ears to lose heat quickly, but arctic foxes have smaller, fatter ears to minimise heat loss.

2. Different materials transfer heat at different rates.
   Example – fur, feathers and human clothing are all poor conductors used to reduce heat loss. In fact, most importantly, they all trap air (a bad conductor), which reduces heat loss even more.
3. Humans sweat and dogs pant.
   Example – as the moisture on our skin, or in a dog's mouth, evaporates it takes heat from its surroundings. This keeps us, and the dog, cool.
4. The surface the object is in contact with.
   Example – when you stand on a carpet in bare feet the floor feels warm because it traps air, which is an insulator. However, a tiled floor is a better conductor so it feels colder even if it is at the same temperature as the carpet.
5. The bigger the temperature difference between an object and its surroundings the faster it transfers heat.
   Example – during a cold winter a house costs more to heat compared to a mild winter. The bigger the temperature difference between inside and outside the quicker the house loses heat.

Examples

A **car radiator** is black, therefore it radiates heat well, and also has a large surface area due to thin cooling fins.

A **vacuum flask** greatly reduces heat flow from the inside to the outside or vice versa. This means that it will keep hot drinks hot or cold drinks cold. A flask is normally made from plastic, which is a poor conductor. The shiny (silvered) sides reflect infrared radiation and stop heat transfer. The vacuum contains no particles, so neither conduction nor convection can take place. The screw top prevents evaporation from the surface and convection currents at the top.

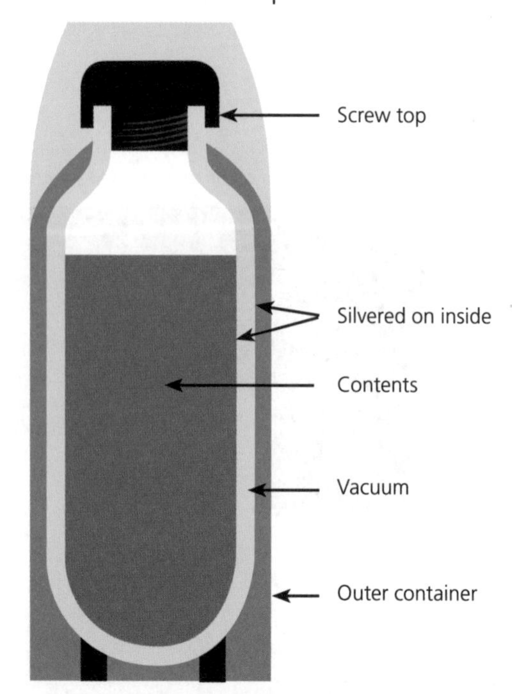

Heating and Insulating Buildings

U-values

The *U*-value of a material indicates how effective a material is as an insulator because it shows how quickly heat energy can pass through. A low *U*-value means that heat flows through it slowly. Therefore, the lower the *U*-value the better the material is at insulating.

Loft insulation | Boiler jacket

Efficiency and Payback Time

When designing houses it is important to consider the *U*-value of the materials used and compare the insulating benefits against the cost of the material. The payback time of a particular improvement tells us how long it would take to make, in efficiency savings, the amount it cost for the improvement. Payback time can be used to work out the cost effectiveness of different types of insulation as well as other improvements (e.g. fitting solar panels).

$$\text{Payback time (years)} = \frac{\text{Total cost of improvement}}{\text{Savings per year}}$$

For example, an improvement that would save £100 per year in heating but costs £20 000 would not be considered cost effective because it has a payback time of 200 years.

Payback time $= \frac{20\,000}{100} =$ **200 years**.

Specific Heat Capacity

The specific heat capacity of a substance is the amount of energy needed to change the temperature of one kilogram of it by one degree Celsius.

$$\text{Energy transferred (J)} = \text{Mass (kg)} \times \text{Specific heat capacity (J/kg°C)} \times \text{Temperature change (°C)}$$

This is useful for deciding which materials to use in heating and cooling applications.

For example, water has a high specific heat capacity, which means it can store a lot of heat energy without getting too warm. This makes it useful as a coolant.

Water is also used in solar heating panels. Water-filled panels placed on a roof absorb heat from the Sun. This radiation energy warms the water. The water in the panel stores a lot of heat energy, which can then be used to heat buildings or provide domestic hot water.

Solar panels are often black and have a large surface area to absorb as much infrared energy as possible.

You need to be able to compare ways in which energy is transferred in and out of objects by heating and ways in which the rate of transfer can be varied.

Every year, people in the UK spend more money on energy (gas and electricity) to heat their homes than they need to. This is because energy gets lost and wasted.

Percentage of Total Heat Loss

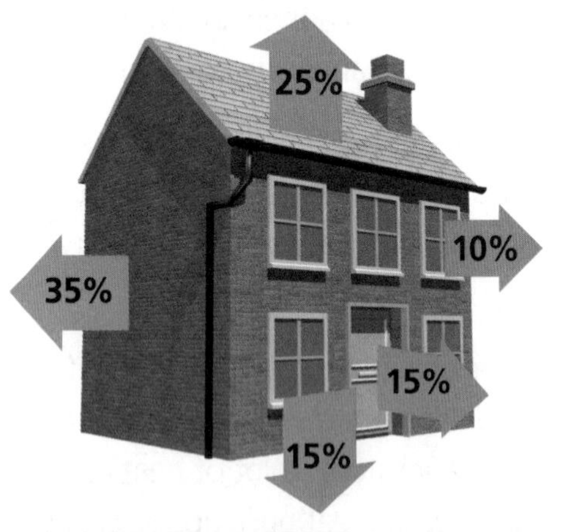

For example, an uninsulated house can lose up to 75% more heat than an insulated house.

Heat energy is transferred from homes into the environment by:

- **conduction** – through the walls, floor, roof and windows
- **convection** – convection currents coupled with cold draughts from gaps in doors and windows cause heat energy carried by warm air to rise up to the roof space where it is easily lost
- **radiation** – from the surface of the walls, roof and through the windows.

Insulating a house will help it to retain the heat in winter, and also help to keep it cool in the summer. The table below outlines how heat can be lost and how this loss can be reduced.

Where Heat is Lost	Preventative Measure	Benefits	Problems
Roof	• Roof insulation – traps a layer of air between fibres or insulating material.	• Can reduce heat loss by 20–25%. • Many different methods to suit all homes. • Short payback time.	• Requires suitable safety precautions to be taken, e.g. wearing dust mask, gloves, etc.
Under doors and windows	• Draught excluders – keep as much warm air inside as possible.	• Can reduce heat loss by up to 15%. • Cheap and easy to install. • Short payback time.	• Must make sure that air vents are not blocked – fresh air needs to circulate to prevent dry rot.
Walls	• Cavity wall insulation and internal thermal boards.	• Can reduce heat loss by 35%.	• Expensive. • Long payback time.
Windows	• Double glazing – traps air between two sheets of glass. • Curtains – stop heat loss through convection.	• Double glazing can reduce heat loss by up to 10%. • Curtains are cheap and easy to install.	• Double glazing is expensive and has a long payback time.
Floor	• Carpets, rugs and underfloor insulation can help to stop heat loss through the floor.	• Carpets and rugs are easy to install.	• Underfloor insulation is expensive and has a long payback time.

P1.2 Energy and efficiency

Energy can be transferred usefully, stored or dissipated, but cannot be created or destroyed. When appliances transfer energy some is always dissipated to the surroundings (wasted). To understand energy efficiency, you need to know:

- how energy is transferred by a range of appliances
- how to calculate efficiency
- how to interpret and draw Sankey diagrams.

Transferring Energy

When devices transfer energy, only part of the energy is usefully transferred to where it is wanted. The rest of the energy is 'wasted'. Wasted energy is eventually transferred to the surroundings, which become warmer. As energy is dissipated to the surroundings it becomes increasingly spread out and so becomes less useful.

Replacing old technology with newer more efficient technology will mean that less of the energy supplied is wasted. When making decisions we need to look at how cost effective the replacement is. The **payback time** of replacing an appliance with a more efficient one depends on how much it costs and how much it saves. When calculating this payback time we use exactly the same method as for insulation.

Efficiency

The greater the proportion of energy that is usefully transferred, the more **efficient** we say device is. For example:

- A car engine is 30% efficient – much more energy is wasted (in heat and sound), than is transferred into useful kinetic energy.
- A microwave oven is 60% efficient – more energy is transferred into useful heat, light and kinetic energy than is wasted (as heat and sound).

Efficiency can be calculated using either energy or power and can be given as a decimal or a percentage. For example, an efficiency of 0.2 is 20% efficient.

$$\textbf{Efficiency} = \frac{\textbf{Useful energy out}}{\textbf{Total energy supplied}} = \frac{\textbf{Useful power out}}{\textbf{Total power supplied}}$$

For example, if a 100W standard light bulb has a light output of 5W, the efficiency would be:

$\frac{5}{100} = \mathbf{0.05}$ **or** multiply by 100 for a percentage = 5%

Energy Efficiency of a Television

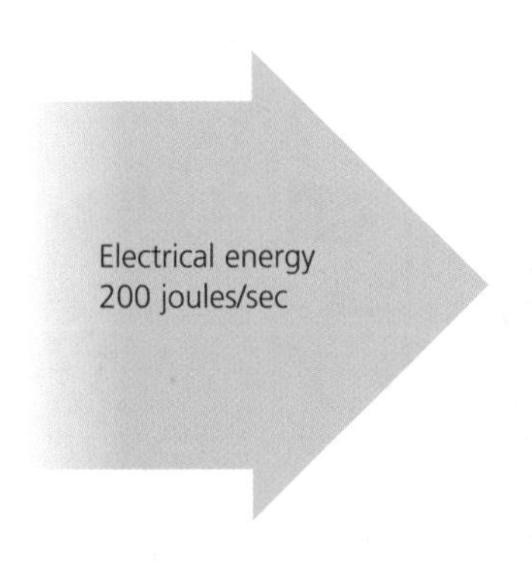

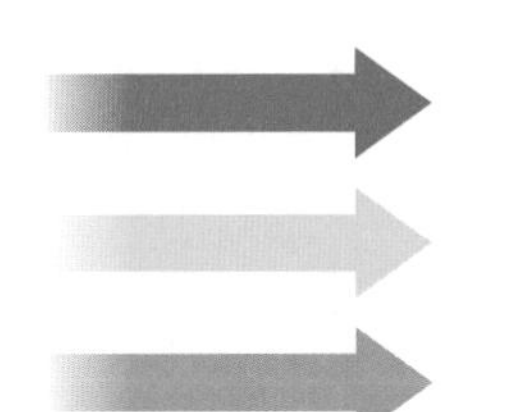

Wasted energy
Heat 150 joules/sec

Useful energy
Light 20 joules/sec

Useful energy
Sound 30 joules/sec

Only a quarter of the energy supplied to a television is usefully transferred into light and sound. Therefore it is only 25% efficient.

and Efficiency

energy
…entation of
…each type. The widths
…proportional to the amount of
…ey represent.

The Sankey diagram below illustrates a standard light bulb. The diagram shows that most of the input electrical energy is wasted as heat. When drawing a Sankey diagram, the useful energy is drawn going straight through and the wasted energy branching off.

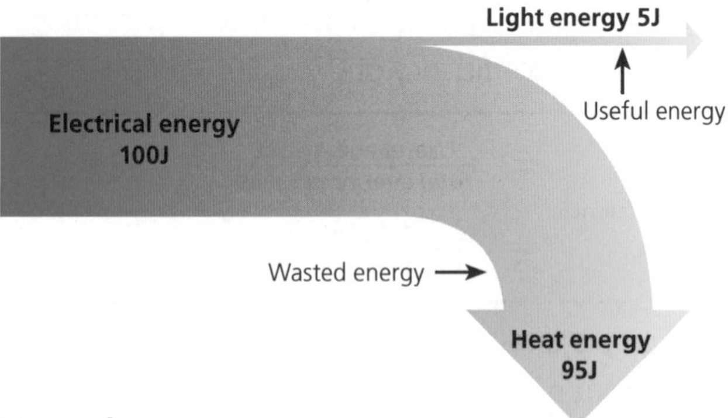

Example

In the example below, the Sankey diagram has been drawn on a square grid. It illustrates the energy transfer that takes place in a car engine, which is 30% efficient.

In the diagram the input energy is 10 squares wide and the useful energy is 3 squares wide.

$\frac{3}{10} = 0.3$ (or 30% efficient)

If the car engine uses 1000J of chemical energy it would usefully transfer 300J to kinetic energy. The rest would be wasted as 500J of heat and 200J of sound.

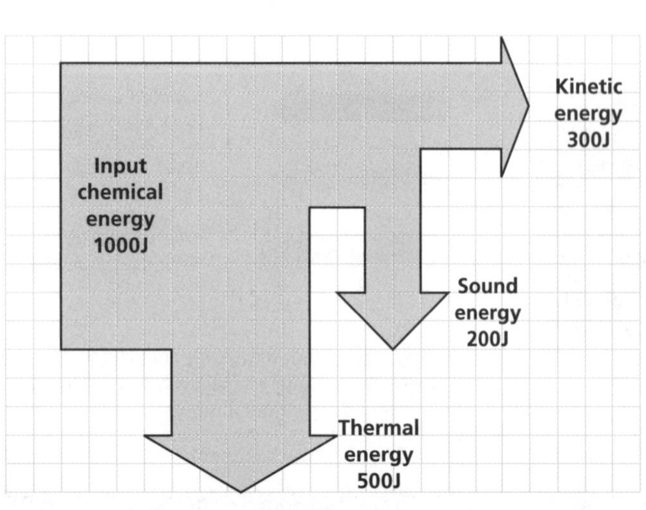

Remember that energy cannot be created or destroyed so when using Sankey diagrams the total energy output (useful + wasted) must **equal** the total energy input.

Example

A microwave oven has an input of 1000J. Draw a Sankey diagram to show the energy transfer of the oven (use a grid where 1 square represents 200J.) How many squares show the useful energy? What is the efficiency of the microwave oven?

The energy of the different outputs are as follows:

Useful: Kinetic – 1 square = 1 × 200 = 200J
Thermal – 1.5 squares = 1.5 × 200 = 300J
Light – 0.5 squares = 0.5 × 200 = 100J

Wasted: Sound – 1 square = 200J
Thermal – 1 square = 200J

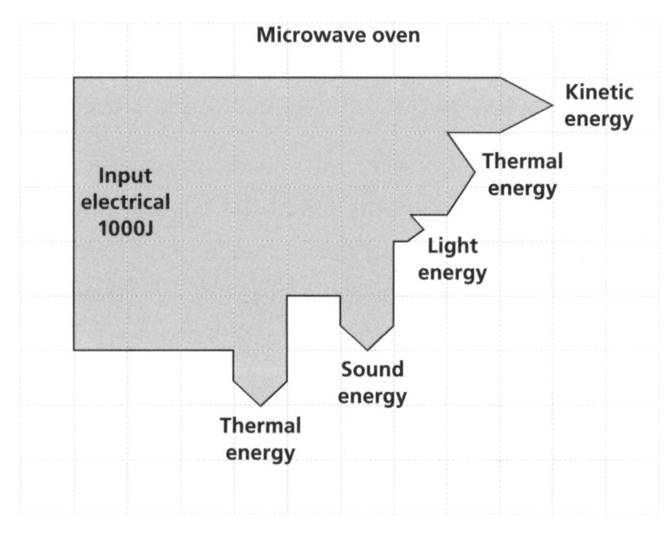

The total useful energy is **3 squares**.

Efficiency = $\frac{3}{5}$ = **0.6** (or 60% efficient)

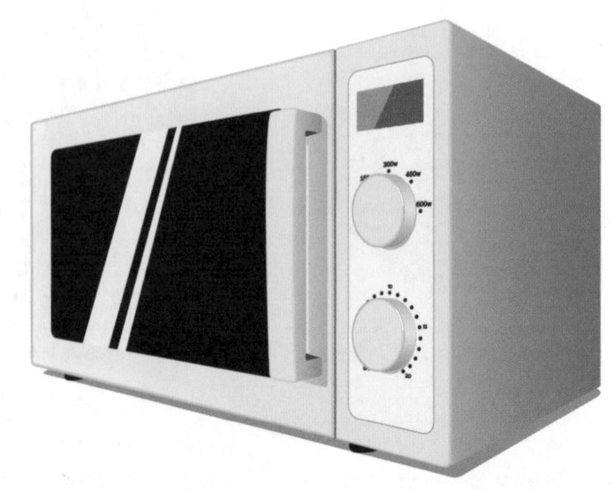

You need to be able to describe the energy transfers and the main energy wastage that occurs with a range of devices, and calculate the efficiency of a device using:

$$\text{Efficiency} = \frac{\text{Useful energy transferred by device}}{\text{Total energy supplied to device}} \times 100$$

Example

Some examples are shown in the table below.

Name of Device and Intended Energy Transfer	Energy In	Energy Out		Efficiency
		Useful	Wasted	
Standard light bulb – electrical energy to light energy	100 joules/sec	Light: 5 joules/sec	Heat: 95 joules/sec	$\frac{5}{100} \times 100\% =$ **5%**
Low energy (compact fluorescent) light bulb – electrical energy to light energy	25 joules/sec	Light: 20 joules/sec	Heat: 5 joules/sec	$\frac{20}{25} \times 100\% =$ **80%**
Kettle – electrical energy to heat energy	2000 joules/sec	Heat (in water): 1800 joules/sec	Heat: 100 joules/sec Sound: 100 joules/sec	$\frac{1800}{2000} \times 100\% =$ **90%**
Electric motor – electrical energy to kinetic energy	500 joules/sec	Kinetic: 300 joules/sec	Heat: 100 joules/sec Sound: 100 joules/sec	$\frac{300}{500} \times 100\% =$ **60%**

You need to be able to evaluate the effectiveness and cost effectiveness of methods used to reduce energy consumption.

Method	Benefits	Problems
Switching lights off when leaving a room	• Easy and simple way to reduce energy consumption.	• People forget to switch lights off or do not like being in a dark house.
Energy-efficient (compact fluorescent) light bulbs	• Use less power so cost less to run. • Less wasted energy. • Last longer than standard bulbs.	• More expensive to buy.
Using electrical equipment during the night, e.g. washing machine	• Cheaper time of day for using electricity (depending on the tariff).	• Can be noisy and keep people awake.
More efficient tumble driers, or letting clothes dry naturally	• Driers with a sensor stop automatically when the clothes are dry, which saves energy.	• New driers are expensive to buy. • Cannot hang clothes out to dry if it is raining
Tankless water heater (combi-boiler)	• Water is heated when needed so less energy is used to heat unnecessary water and keep it heated.	• Some units do not have enough power to supply to more than one tap at a time.

P1 The Usefulness of Electrical Appliances

P1.3 The usefulness of electrical appliances

Electrical energy can be transferred easily across large distances and therefore is very suitable to use in our homes and industries. To understand this, you need to know:

- what energy transformations electrical devices bring about
- how energy and power are measured

Energy Transformation

Most of the energy transferred to homes and industry is electrical energy. Electrical energy is easily transformed to:

- heat (thermal) energy, e.g. an electric fire
- light energy, e.g. a lamp
- sound energy, e.g. stereo speakers
- movement (kinetic) energy, e.g. an electric whisk.

The **power** of an appliance is measured in **watts** (**W**) or **kilowatts** (**kW**). **Energy** is normally measured in **joules** (**J**). *N.B. 1W = 1J/s*

$$\textbf{Power}\ (\text{W}) = \frac{\textbf{Energy}\ (\text{J})}{\textbf{Time}\ (\text{s})}$$

Calculating the Amount and Cost of Energy Transferred

The amount of energy transferred from the mains can be calculated using the equation:

$$\underset{(\text{kilowatt-hour, kWh})}{\textbf{Energy transferred}} = \underset{(\text{kilowatt, kW})}{\textbf{Power}} \times \underset{(\text{hour, h})}{\textbf{Time}}$$

The **cost of energy** transferred from the mains can be calculated using the equation:

$$\textbf{Total cost} = \textbf{Number of kilowatt hours} \times \textbf{Cost per kilowatt hour}$$

REB Regional **Electricity** Board

Mr R. Jones
273 Dove Street
Southampton
SW15 WFK

Electricity Statement. Period: 01.01.11 – 01.04.11 No standing charge

Present reading	Previous reading	kWh used	Cost per kWh (p)	Charge amount (£)
12898 (economy)	12640 (economy)	258	7.32	18.89
30803	30332	471	9.45	44.51
			Total VAT exclusive charges	63.40
			VAT at 5%	3.17
			Total charges including VAT	66.57

Regional Electricity Board, Anchor House, Ingleby Street, Southampton SW15 TNE **Telephone:** 01445 680180 **Fax:** 01445 680180 **Email:** info@reb.co.uk **Web:** www.reb.co.uk

Economy reading – electricity used during the night

Normal reading – electricity used during the day

kWh calculated by subtracting present reading from previous reading

Kilowatt-hours is the unit of electricity

Cost
= kWh x cost per kWh
= 258 x 7.32p

Total charge
= economy charge + normal charge

VAT = 63.40 x $\frac{5}{100}$

You need to be able to compare and contrast different electrical devices that can be used for a particular application, and calculate the amount of energy transferred from the mains using this equation:

Energy transferred (kilowatt-hour, kWh)	=	**Power** (kilowatt, kW)	×	**Time** (hour, h)

Example

Mr and Mrs Jones are shopping for a new shower for their house. They see three different showers advertised (see below).

They work out the advantages and disadvantages of each model in order to decide which one will be the most suitable for their needs. They also need to work out how much energy will be needed to power each one. They have calculated that every morning they each have a shower lasting, on average, 12 minutes, while their young son, Nick, has a shorter shower lasting 6 minutes (total time = 12 + 12 + 6 = 30 min or 0.5 hours).

Using the formula they calculate how much energy will need to be transferred from the mains to power each shower:

Power × Time = Energy

7kW: $7 \times 0.5 = 3.5$kWh

8.5kW: $8.5 \times 0.5 = 4.25$kWh

9.5kW: $9.5 \times 0.5 = 4.75$kWh

To compare the showers they work out the following:

- The 7kW shower will require the least amount of energy from the mains, but it has a less powerful spray and fewer options.
- The 8.5kW shower will require more energy than the 7kW shower, but it has a more powerful spray.
- The 9.5kW shower is more expensive to buy and requires more energy, but it has a much more powerful spray, more functions and can be connected to a cold water tank. This will be more useful for a family because they won't have to wait for the water to heat up.

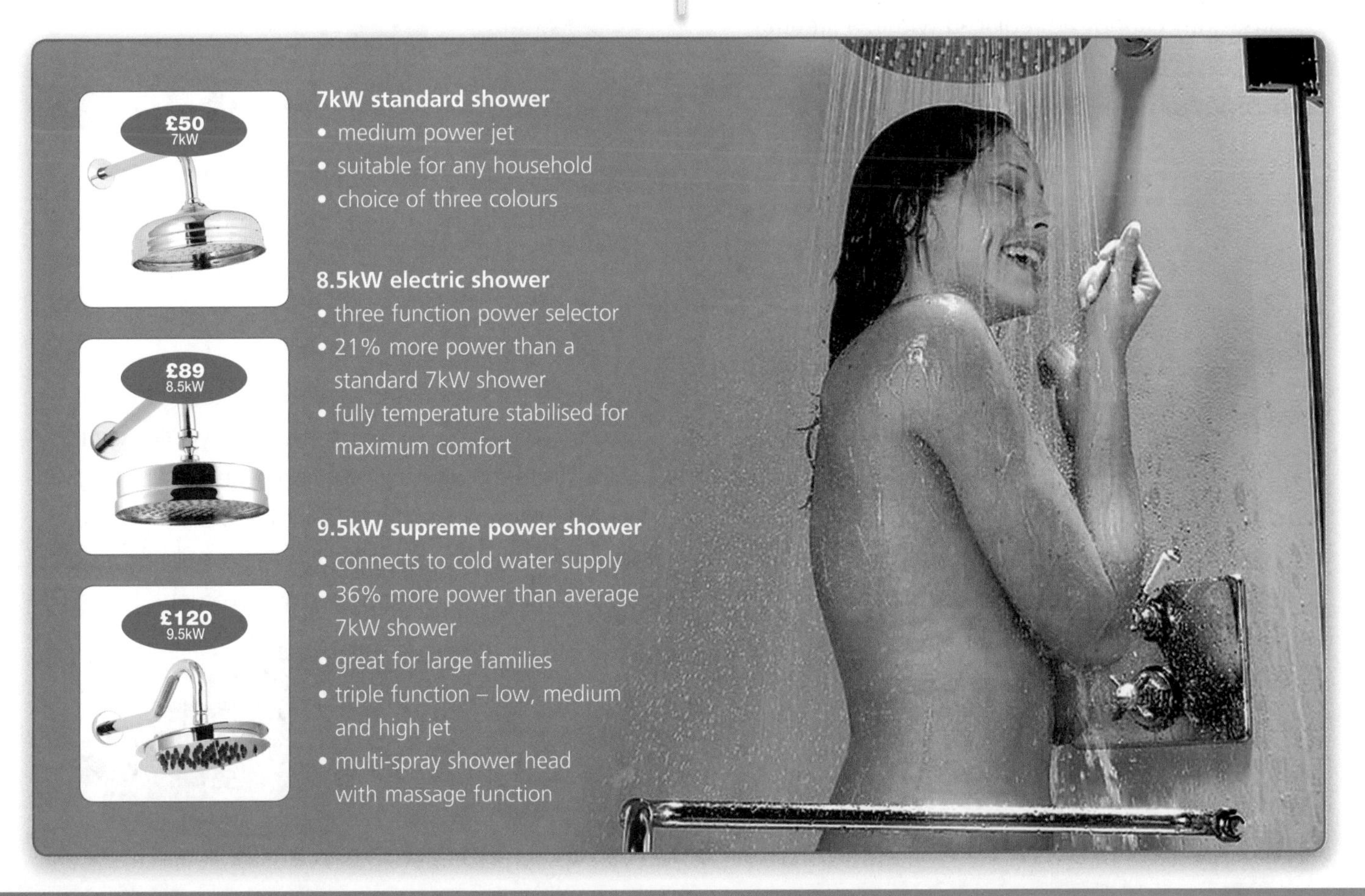

P1.4 Methods we use to generate electricity

Various energy sources can be used to generate the electricity that we need. We need to understand the benefits and costs associated with each method to select the best one to use in a particular situation. To understand this, you need to know:

- how energy is produced
- about renewable and non-renewable energy sources
- the advantages and disadvantages of using different energy sources
- how electricity is distributed, including the role of transformers.

Non-renewable Energy Sources

Coal, oil and gas are energy sources that formed over millions of years from the remains of plants and animals. They are called **fossil fuels** and are used to produce most of the energy that we use. However, because they cannot be replaced within a lifetime they will eventually run out. They are therefore called **non-renewable** energy sources.

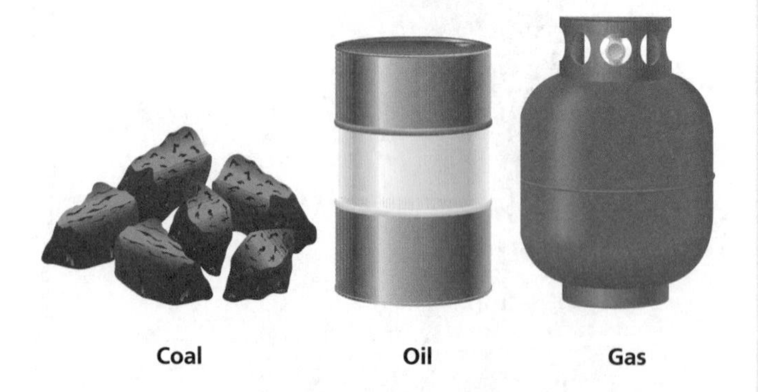

Nuclear fuels, such as uranium and plutonium, are also non-renewable. Nuclear fission is the splitting of a nucleus releasing neutrons that collide with other nuclei causing a chain reaction that generates huge amounts of heat energy. However, nuclear fuel is not burnt (like coal, oil or gas) to release energy and is not classed as a fossil fuel.

Generating Electricity from Non-renewable Sources

In power stations, fossil fuels are burned to release heat energy, which boils water to produce steam. The steam is used to drive turbines, which are attached to electrical generators. A gas jet turbine generator heats air instead of water giving it a quicker start-up time.

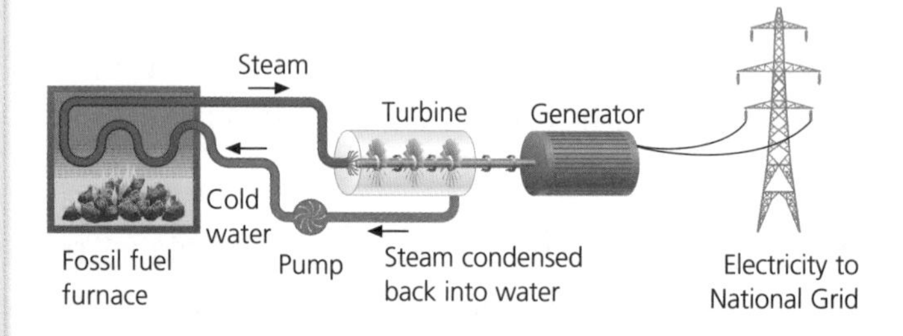

Nuclear fuel is used to generate electricity in a similar way. A reactor is used to generate heat by nuclear fission. A heat exchanger is used to transfer the heat energy from the reactor to the water, which turns to steam and drives the turbines.

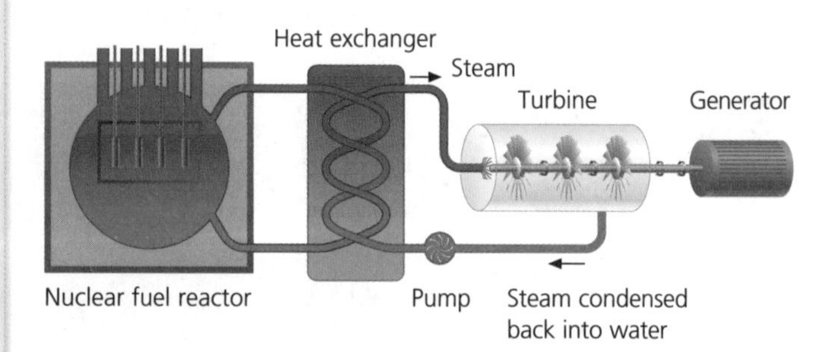

Biofuels

Biofuels – e.g. wood and rapeseed oil (used for bio-diesel) – are also burnt to release energy. Because the plants used grow relatively quickly, the fuels burned can be replaced. For this reason biofuels are classed as **renewable** energy sources. Plant and animal waste from farming can also be used as a biofuel.

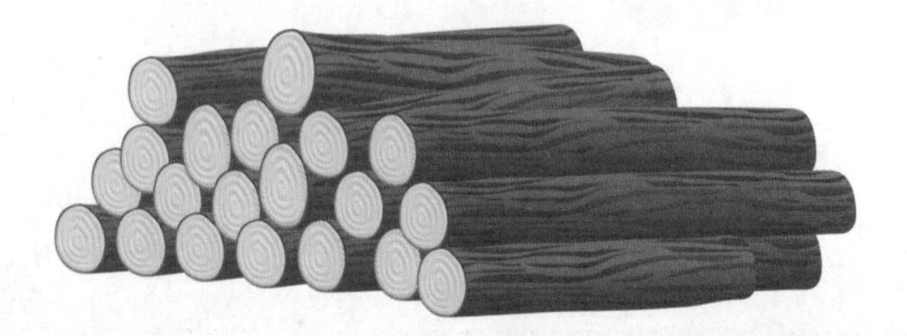

Methods We Use to Generate Electricity

Comparing Non-renewable Sources of Energy

The energy sources below are used to provide most of the electricity we need in this country through power stations. Some of the advantages and disadvantages of each are listed below.

Source	Advantages	Disadvantages
Coal	• Relatively cheap and easy to obtain. • Coal-fired power stations are flexible in meeting demand and have a quicker start-up time than their nuclear equivalents. • Estimates suggest that there may be over a century's worth of coal left.	• Burning produces carbon dioxide (CO_2) and sulfur dioxide (SO_2). • Produces more CO_2 per unit of energy than oil or gas does. (CO_2 causes global warming.) • SO_2 causes acid rain unless the sulfur is removed before burning or the SO_2 is removed from the waste gases. Both of these add to the cost.
Oil	• Enough oil left for the short to medium term. • Relatively easy to find, though the price is variable. • Oil-fired power stations are flexible in meeting demand and have a quicker start-up time than both nuclear-powered and coal-fired reactors.	• Burning produces CO_2 and SO_2. (CO_2 causes global warming, SO_2 causes acid rain.) • Produces more CO_2 than gas per unit of energy. • Often carried between continents on tankers leading to the risk of spillage and pollution.
Gas	• Enough natural gas left for the short to medium term. • Can be found as easily as oil. • No SO_2 is produced. • Gas-fired power stations are flexible in meeting demand and have a quicker start-up time than nuclear, coal and oil-fired reactors.	• Burning produces CO_2, although it produces less than coal and oil per unit of energy. (CO_2 causes global warming.) • Expensive pipelines and networks are often required to transport it to the point of use.
Nuclear	• Cost and rate of fuel consumption is relatively low. • Can be situated in sparsely populated areas. • Nuclear power stations are flexible in meeting demand. • No CO_2 or SO_2 produced.	• Although there is very little escape of radioactive material in normal use, radioactive waste can stay dangerously radioactive for thousands of years and safe storage is expensive. • Building and decommissioning is costly. • Longest comparative start-up time.

Summary of Non-renewable Sources of Energy

Advantages	Disadvantages
• Produce huge amounts of energy. • Reliable. • Flexible in meeting demand. • Do not take up much space (relatively).	• Pollute the environment. • Cause global warming and acid rain. • Will eventually run out. • Fuels often have to be transported over long distances.

P1 Methods We Use to Generate Electricity

Renewable Energy Sources

Renewable energy sources are those that will not run out because they are continually being replaced. Many of them are caused by the Sun or Moon. The gravitational pull of the Moon creates tides. The Sun causes:

- evaporation, which results in rain and flowing water
- convection currents, which result in winds, which in turn create waves.

Generating Electricity from Renewable Energy Sources

Renewable energy sources can be used to drive turbines or generators directly. In other words, no fuel needs to be burnt to produce heat.

The table below shows the most common methods of generating energy from renewable energy sources.

Wind Turbines

Wind can be used to drive huge turbines, which, in turn, drive generators. Wind turbines are usually positioned on the top of hills so they are exposed to as much wind as possible.

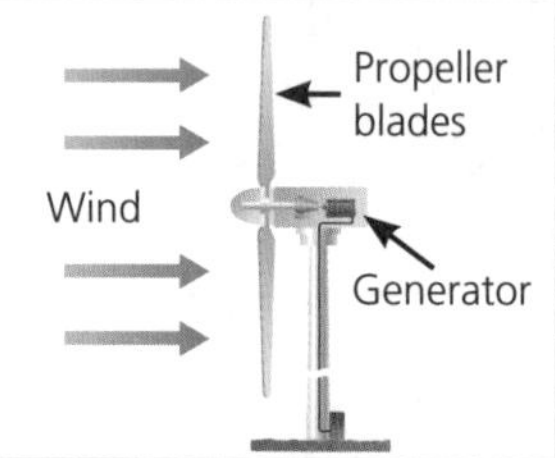

Solar Cells and Panels

Solar cells and panels are made of a semiconductor material (usually silicon), which captures the light energy and transforms it into electrical energy.

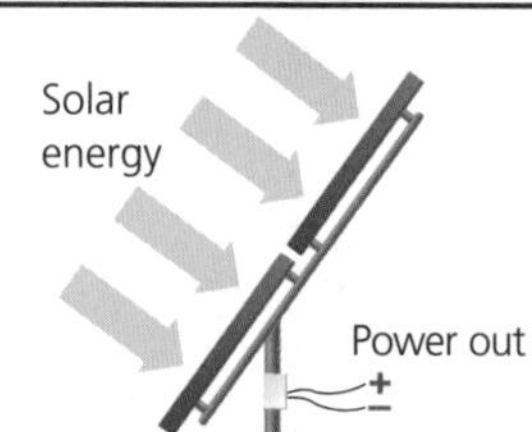

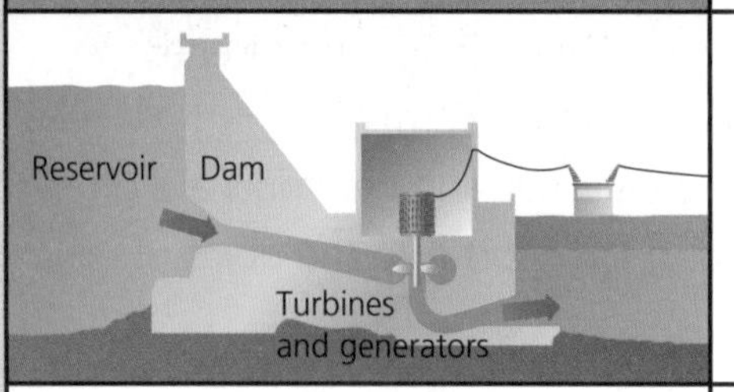

Hydroelectric Dam

Water that is stored in a reservoir above the power station is allowed to flow down through the pipes to drive the generator, which produces a lot of power.

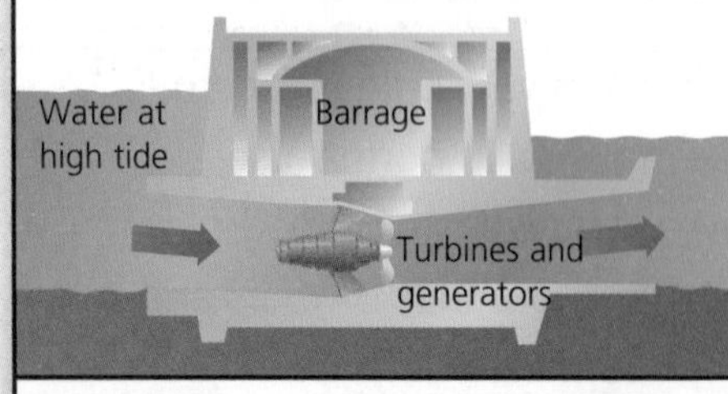

Tidal Barrage

As the tide comes in water flows freely through a valve in the barrage. This water then becomes trapped. At low tide, the water is released from behind the barrage through a gap that has a turbine in it. This drives a generator.

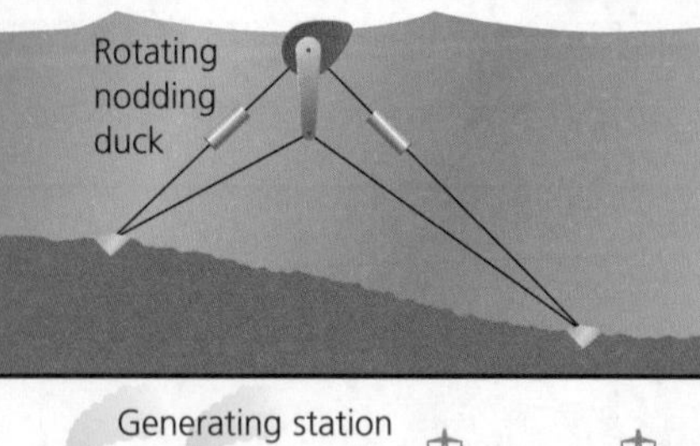

Nodding Duck

Nodding ducks are found in the sea. The motion of the waves makes the 'ducks' rock and this movement is translated into a rotary movement which, in turn, drives a generator.

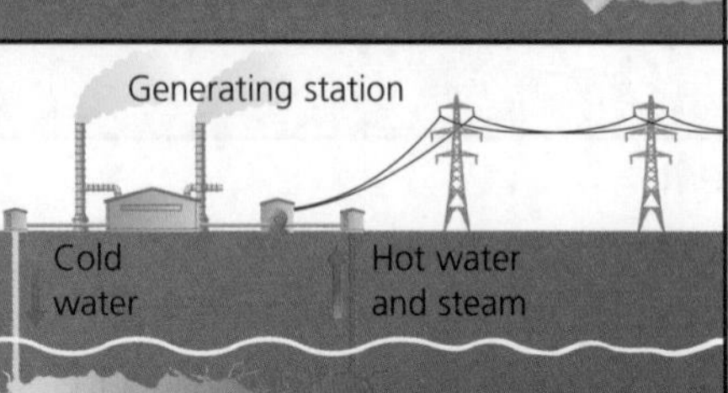

Geothermal

In some volcanic areas hot water and steam rise naturally to the surface, having been heated up by the decay of radioactive substances (e.g. uranium) within the Earth. This steam can be used directly to drive turbines which, in turn, drive generators.

Methods We Use to Generate Electricity

Comparing Renewable Sources of Energy

The energy sources listed below use modern technology to provide us with a clean, safe alternative source of energy. Some of their advantages and disadvantages are given.

Source	Advantages	Disadvantages
Wind	• No fuel and little maintenance required. • No pollutant gases produced. • Once built they provide 'free' energy when the wind is blowing. • Can be built offshore.	• Need many turbines to produce a sizable amount of electricity, which means noise and visual pollution. • Electricity output depends on the strength of the wind. • Not very flexible in meeting demand unless the energy is stored. • Building cost can be high.
Tidal and Wave	• No fuel required. • No pollutant gases produced. • Once built they provide 'free', reliable energy. • Barrage water can be released when demand for electricity is high.	• Tidal barrages, across estuaries, are unsightly, a hazard to shipping and destroy the habitats of wading birds, etc. • Daily variations of tides and waves affect output. • High initial building cost.
Hydro-electric	• No fuel required unless storing energy to meet future demand. • Fast start-up time to meet sudden demand. • Produce large amounts of clean, reliable electricity. • No pollutant gases produced. • Water can be pumped back up to the reservoir when demand for electricity is low, e.g. in the night.	• Location is critical and often involves damming upland valleys, which means flooding farms, forests and natural habitats. • To achieve a net output (aside from pumping) there must be adequate rainfall in the region where the reservoir is. • Very high initial capital outlay (though worth the investment in the end).
Solar	• Ideal for producing electricity in remote locations. • Excellent energy source for small amounts. • Produces free, clean electricity. • No pollutant gases produced.	• Dependent on the intensity of light; more useful in sunny places. • High cost per unit of electricity compared to all other sources, except non-rechargeable batteries.

Summary of Renewable Sources of Energy

Advantages	Disadvantages
• No fuel costs during operation. • No chemical pollution. • Often low maintenance. • Do not contribute to global warming or acid rain formation.	• With the exception of hydroelectric and tidal, they produce small amounts of electricity. • Take up lots of space and are unsightly. • Unreliable (apart from hydroelectric and tidal), depend on the weather and cannot guarantee supply on demand. • High initial cost.

New Technologies – Small-Scale Production

As renewable technologies improve, small-scale production is becoming more cost effective. Solar panels are used for road signs and lights in remote areas that are not connected to the National Grid.

Solar panels, wind turbines, ground pump heating and even micro hydroelectric systems are now available for private homes. Installing these systems can save a home owner hundreds of pounds a year in electricity and heating bills. The payback time is 10–25 years depending on the method used.

Carbon Capture

Carbon capture and storage is a rapidly growing technology being used to deal with the carbon dioxide produced by burning fossil fuels. Some of the best storage containers are oil and gas fields.

Carbon dioxide can be pumped into gas fields, displacing the natural gas that is burned as fuel. Carbon dioxide can also be pumped into oil fields, which keeps the oil under pressure making it easier to extract.

Solar Furnaces

Large-scale solar production uses hundreds of mirrors to reflect infrared energy from the Sun on to a tank full of water.

This energy boils the water, which produces steam that is used for driving turbines in the same way as in a normal power station. An advantage is that no polluting gases are released.

By 2013 the Solucar complex in southern Spain will be producing enough energy for 240 000 homes.

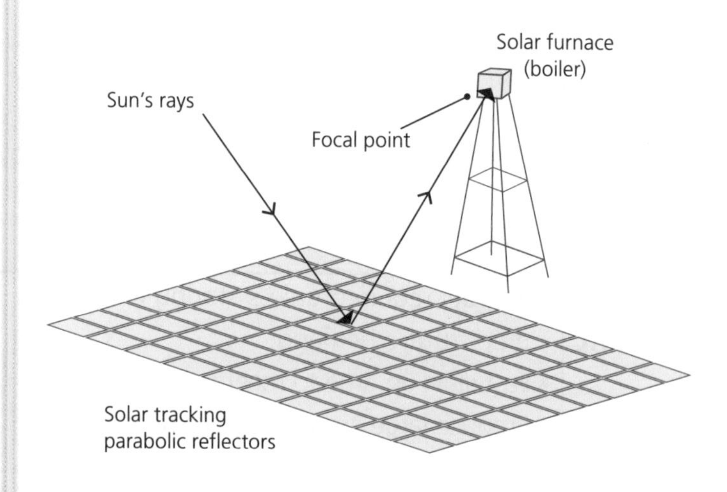

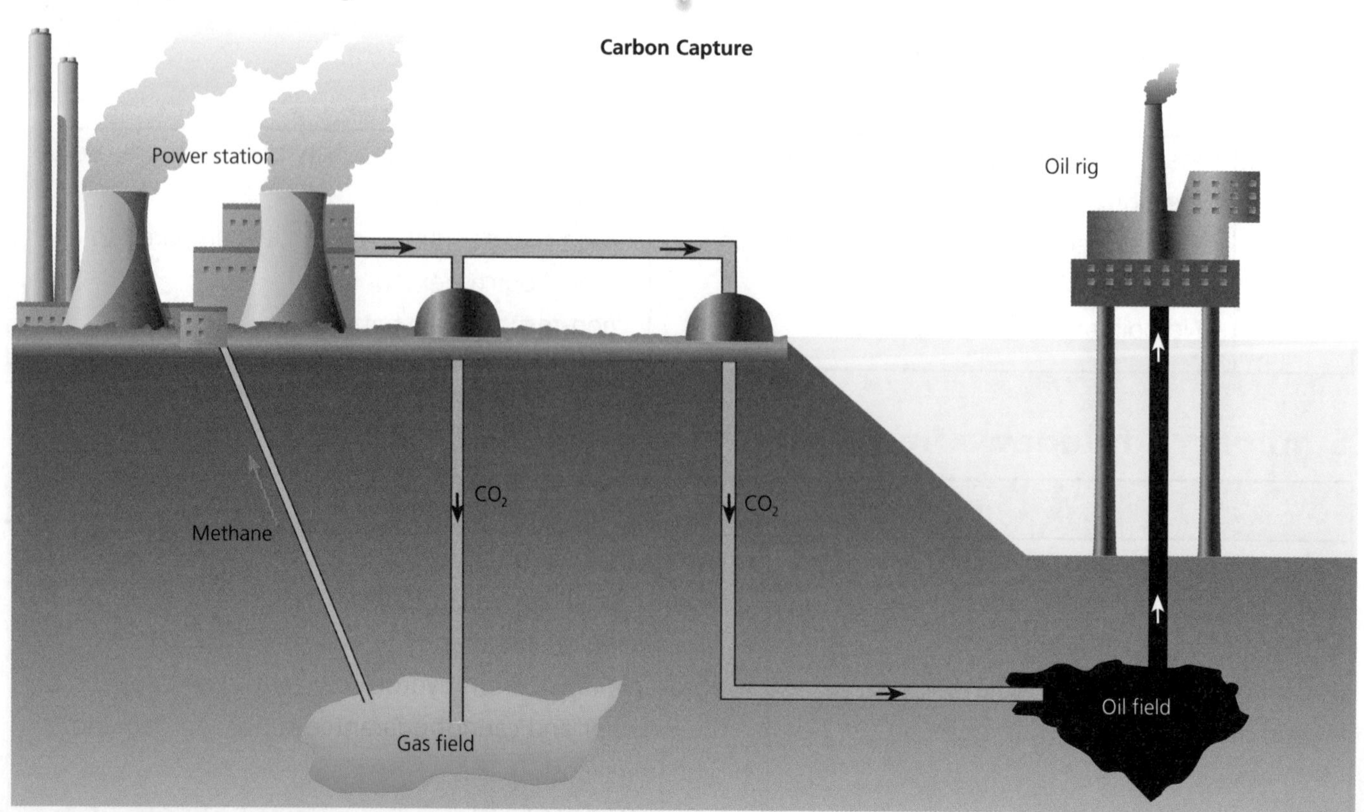

The National Grid

Electricity generated at power stations is distributed to homes, schools and factories all over the country by a network of cables called the **National Grid**. **Transformers** are used to change the **voltage** of the alternating current supply before and after it is **transmitted** through the National Grid. Where it is economically viable to do so, small-scale production systems also provide electricity to the grid.

Overhead power lines are used in the countryside – they are cheaper to install and can cross roads and rivers easily. **Underground** cables are used in towns because of the tall buildings and for safety (people can't access them).

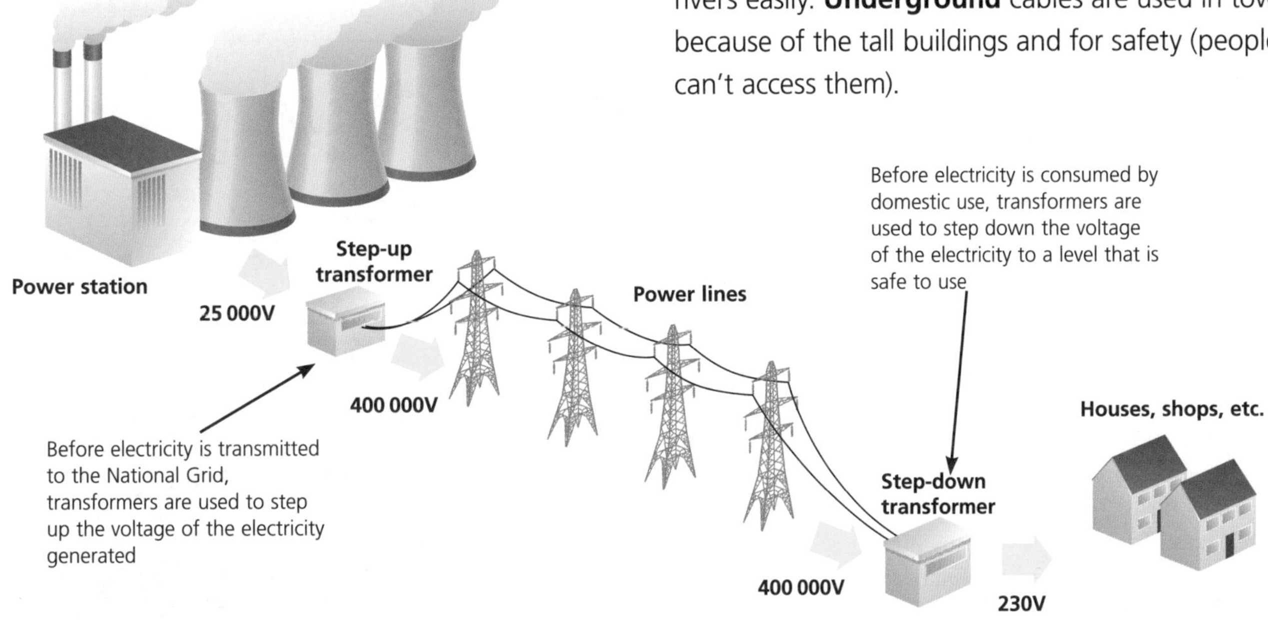

Reducing Energy Loss During Transmission

For a given power, increasing the voltage reduces the current. When a step-up transformer increases the voltage, it reduces the current. The lower the current that passes through a wire the cooler the wire stays and the less energy is wasted as heat. Therefore, electricity needs to be transmitted at as low a current as possible.

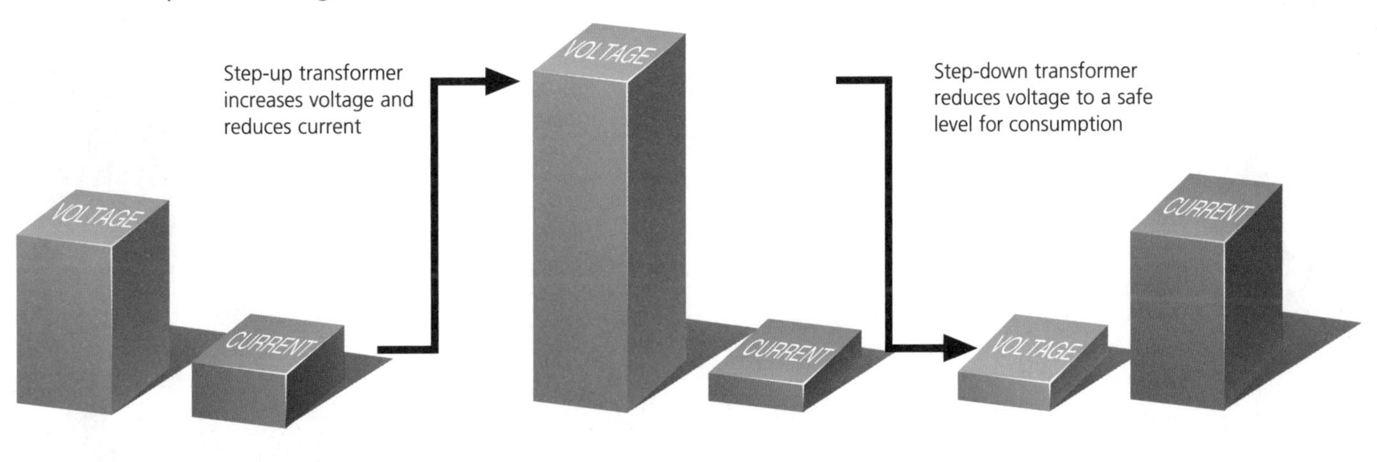

Power stations need to transmit electricity with a high voltage and low current

Power lines – electricity is transmitted at high voltage but low current to reduce the amount of energy lost

For **domestic use** electricity needs to be at a lower voltage and higher current

You need to be able to compare and contrast the particular advantages and disadvantages of using different energy sources to generate electricity and consider what would happen if electricity was not available.

Example

The Lonsdale News

Power to the People

The little village of Bukestead was in uproar yesterday after proposed plans were unveiled for a nuclear power station to be built just 10 miles away from the village on the top of the moor. The local area is well known for its beautiful countryside and is a carefully controlled breeding area for many birds. However, increasing demand for electricity means that new electricity generating plants need to be built and this site is considered to be most suitable. Its isolated location and the large number of jobs it will create, have made nuclear power a popular choice with some people.

However, opinion is strongly divided as to the best use of the land, with renewable energy options being cited by many as a better alternative.

Join the debate. Write to your local MP or councillor and have your say.

6 Moorland Road
Bukestead
Derbyshire

Mrs J Smith (MP)
Derby Council Offices
134 High Street
Derby

Dear Mrs Smith

I was appalled to read about the proposed nuclear power station. The blight on the countryside, the traffic and the horror of a nuclear leak would be unthinkable! What would happen to the local countryside, farms, and the breeding grounds for migrating birds that have been so carefully built up? The local roads would be constantly clogged up with tankers travelling to and from the plant, some carrying highly dangerous waste for disposal. Not to mention the expense of building the site!

This land would be ideal for a farm of wind turbines. The constant strong winds on the hill, coupled with adequate space for 10 to 15 turbines, would be sufficient to produce a sizable amount of electricity. There would be no pollution, no fuel and little maintenance required, just clean, free energy. I hope you will consider the alternatives.

Yours sincerely,

Anne Pentwhistle

(A very concerned local resident)

26 Hill View, Bukestead, Derbyshire

Mrs J Smith (MP)
Derby Council Offices
134 High Street
Derby

Dear Mrs Smith

I am writing in support of the proposed nuclear power station.

The isolated position on the hill would be ideal. Nuclear leaks are so uncommon that no real threat would be posed and the local economy would greatly benefit from the number of jobs created.

I know that reasons are being cited for the use of 'green' electricity – big solar panels, huge wind turbine farms and the like. These are unfeasible solutions; they could never fulfil the demand, and they themselves would visually pollute the countryside. Everyone says, 'not on my door step', but nuclear is flexible, non-pollutant and will provide the electricity needed for the area.

The need for reliable electricity is very important. Without it we could have endless power cuts, creating the need to use candles for lighting, and leaving some people without heating or cooking facilities. Even worse it could cause accidents when road lights stop working and cause problems in hospitals.

I whole-heartedly support your plans.

Yours sincerely,

A. Pollard

Local resident

P1.5 The use of waves for communication and to provide evidence that the Universe is expanding

All waves transfer energy without transferring matter. Waves can be either transverse or longitudinal. To understand this, you need to know:

- about reflection, refraction and diffraction
- the different types of waves
- about speed, frequency, wavelength and amplitude
- the wave equation.

Wavelength, Frequency and Amplitude

The diagram shows a wave. The **wavelength** is the length of one complete wave, measured in metres (m). The **amplitude** is the height from the middle of the wave to the top.

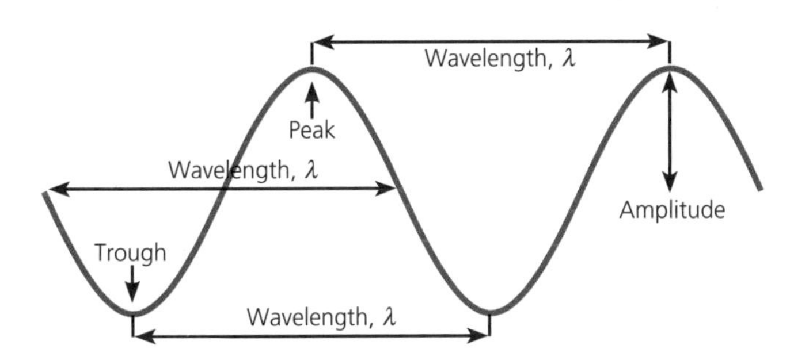

There are two other properties of waves not shown on the diagram:

- **Wave speed** – how quickly the energy moves, measured in metres per second (m/s).
- **Frequency** – how many waves pass a fixed point in a second, measured in hertz (Hz).

All waves obey the wave equation:

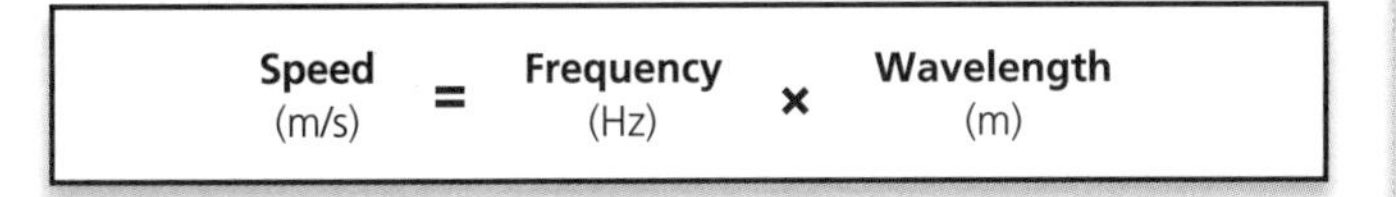

Transverse Waves

Waves in water are an example of a transverse wave. The water moves up and down as the wave travels along.

We say that the oscillations are perpendicular to the direction of energy transfer.

A slinky spring can be used to show transverse waves.

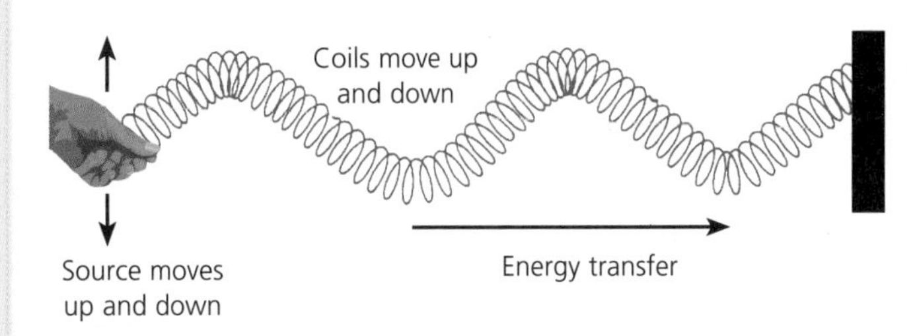

Longitudinal Waves

A slinky spring can also show longitudinal waves. In a longitudinal wave the oscillations are parallel to the direction of energy transfer.

Places where the particles are bunched together are called **compressions**; where they are spread apart they are called **rarefactions**.

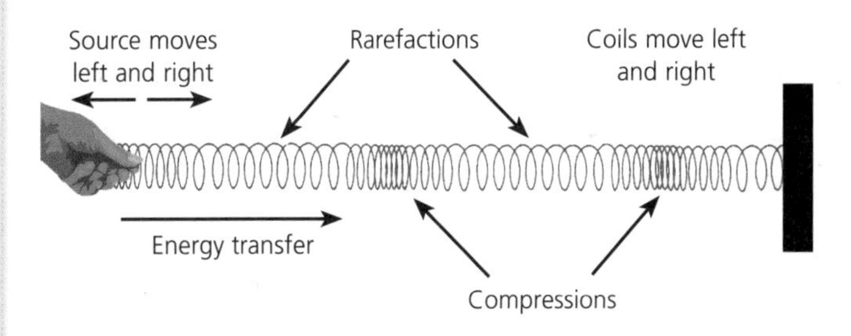

Reflection

When waves are incident on a surface they can be reflected, transmitted or absorbed. An echo is heard when sound is reflected. When light is reflected an image can be formed.

Reflection of Light

When light rebounds off a surface and changes direction it is **reflected**. The diagram shows light being reflected in a plane mirror (flat, smooth, shiny).

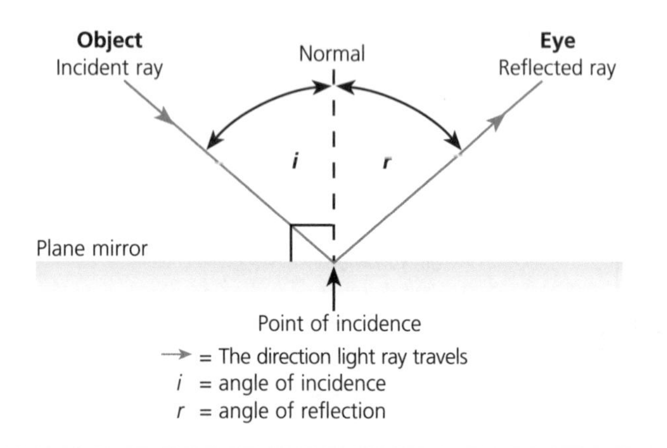

Angle of incidence = Angle of reflection

The **normal line** is perpendicular to the reflecting surface at the point of incidence. The normal is used to calculate the angles of incidence and reflection. The **incident ray** is the light ray travelling towards the mirror. The **reflected ray** is the light ray travelling away from the mirror.

Images Produced by Mirrors

The image formed by a **plane** mirror is the same size as the object. It is upright and laterally inverted (faces the opposite way to the object). The rays of light reaching the eye appear to come directly from the image. Because the rays of light do not actually come from the image, it is described as a **virtual** image.

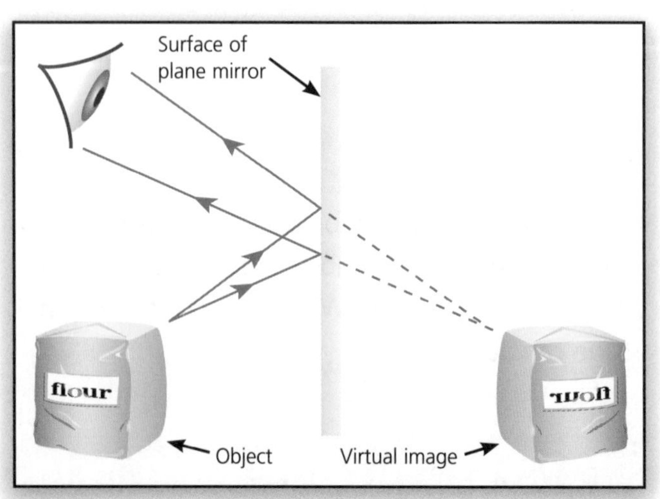

Ray Diagrams

You need to be able to draw ray diagrams illustrating images formed by a plane mirror. By following these steps you should be able to do so accurately. The green arrow shows the object.

1. First draw the image exactly opposite the object the same distance from the mirror.

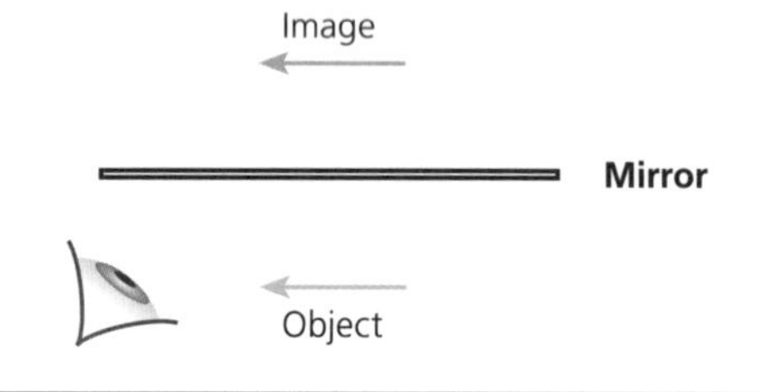

2. Then draw a straight line from one end of the image to the eye. The line behind the mirror should be a dotted line.

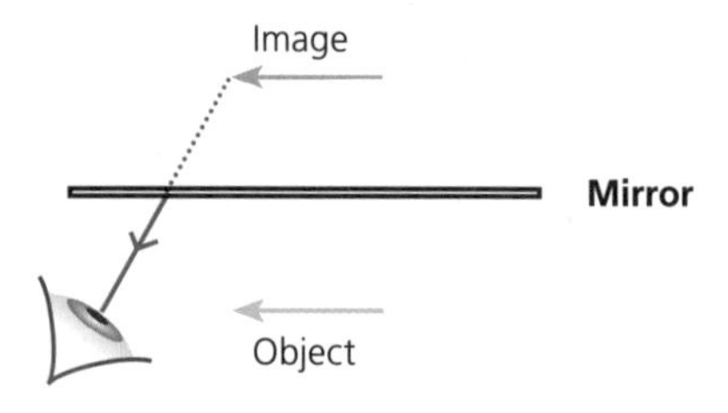

3. Add a line from the corresponding end of the object to the point where the previous line met the mirror. This now shows the ray travelling from the object, reflecting off the mirror and entering the eye. It can be seen that the law of reflection is being obeyed.

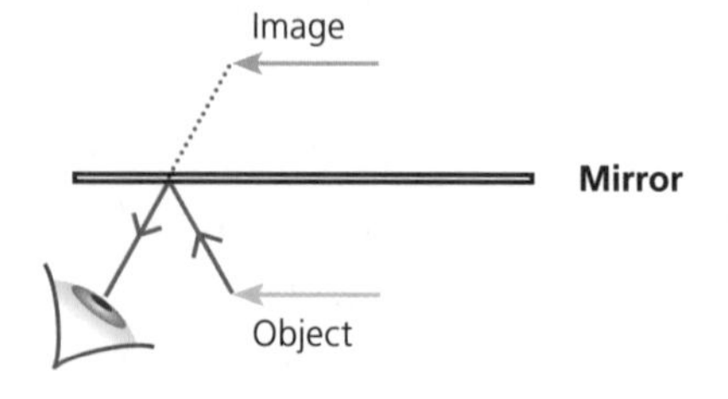

4. Repeat for the other end of the image.

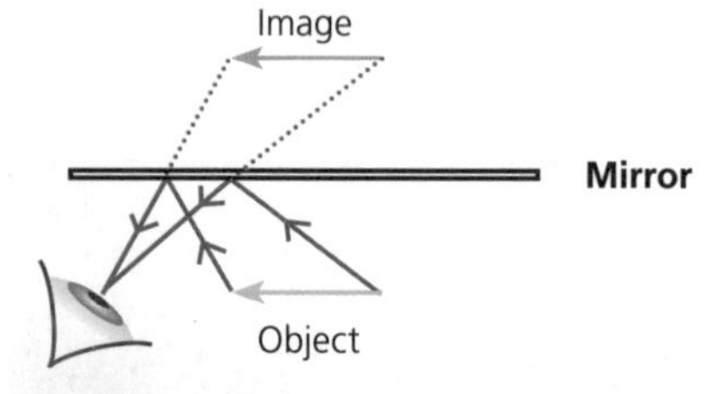

Refraction and Diffraction

As well as being reflected, waves can be **refracted** and **diffracted**.

Refraction occurs when a wave crosses an interface between different substances. Waves travel at different speeds in different mediums, e.g. water waves moving from deep to shallow water or light passing from air to glass.

Refraction of Light at an Interface

Light changes direction when it crosses an interface, i.e. a boundary between two transparent materials (media) of different densities. If a light ray meets the boundary at an angle of 90° (i.e. along the normal) then the direction remains unchanged.

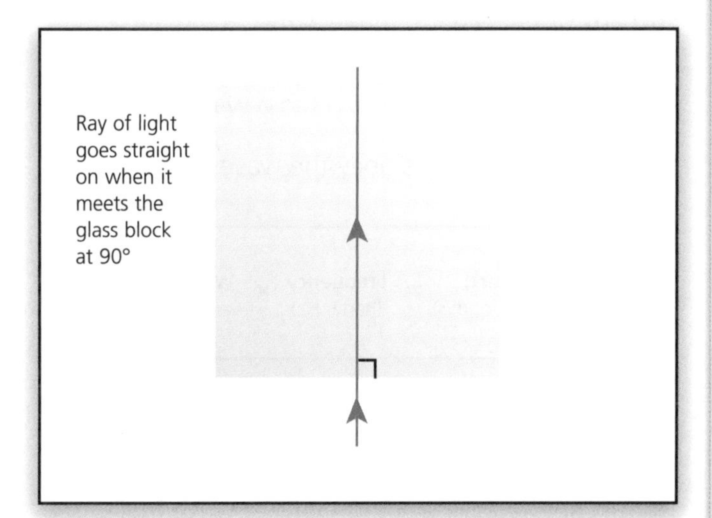

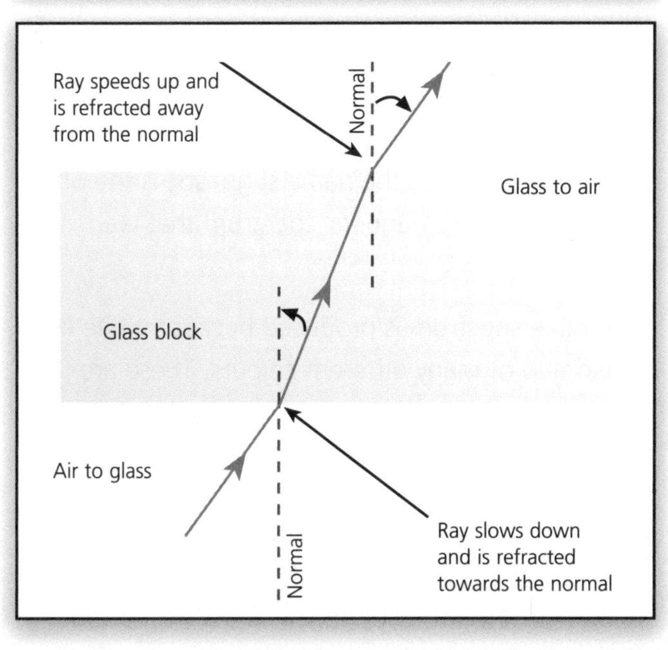

Diffraction

Diffraction occurs when a wave passes through a narrow opening. This can be seen when waves enter a harbour.

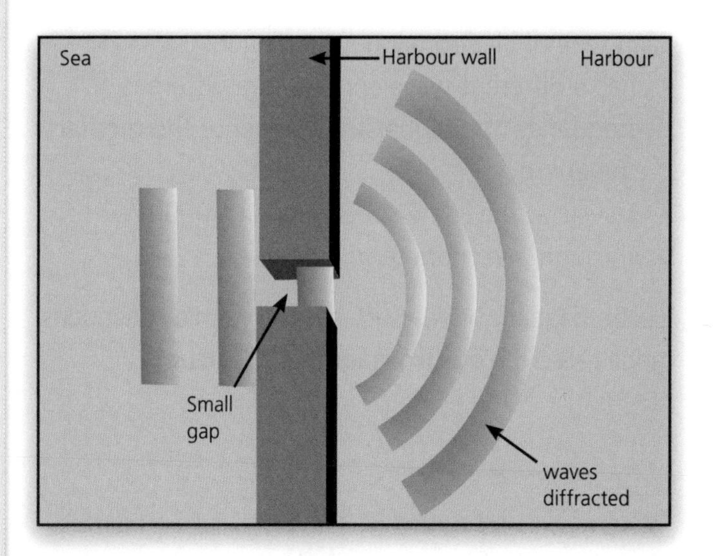

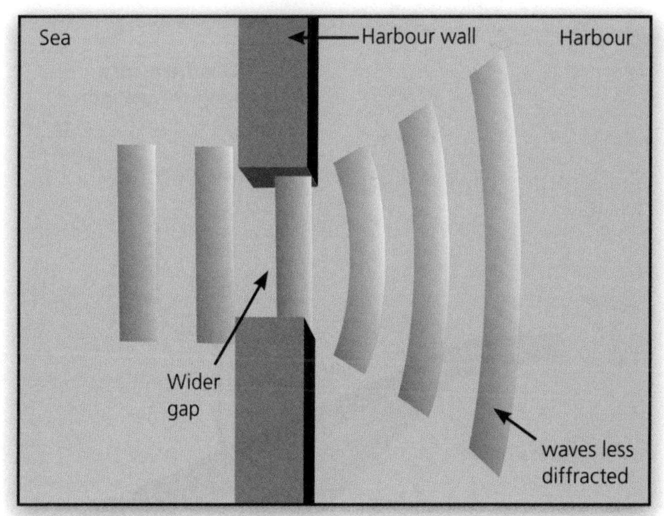

Diffraction is greatest when the gap is the same width as the wavelength. For this reason we do not often see diffraction of light, but it can be seen producing colours on a CD.

Electromagnetic Waves

Electromagnetic waves are transverse waves – they are the only type of wave that can travel through a vacuum.

Each type of electromagnetic radiation:

- has a different wavelength and a different frequency (the higher the frequency the greater the energy)
- travels at the same speed (300 000 000m/s) through a vacuum (e.g. space).

Electromagnetic waves (such as light) form a continuous range called the **electromagnetic spectrum**.

Low frequency, long wavelength
Radio waves
Microwaves
White light
Infrared rays
Visible light
Glass prism
Ultraviolet rays
X-rays
Gamma rays
High frequency, short wavelength

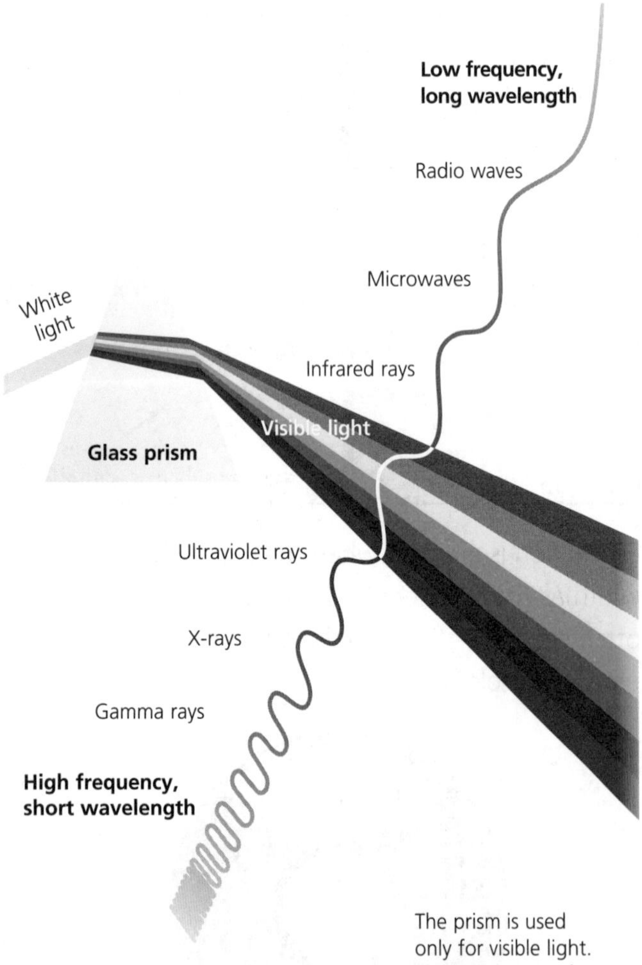

The prism is used only for visible light. The other parts of the electromagnetic spectrum are shown for completion.

Radio waves, microwaves and infrared rays all have a longer wavelength and a lower frequency than visible light.

Ultraviolet waves, X-rays and gamma rays all have a shorter wavelength and a higher frequency than visible light.

Electromagnetic waves can be **reflected** and **refracted**. Different wavelengths of electromagnetic waves are reflected, absorbed or transmitted differently by different substances and types of surface, e.g. black surfaces are particularly good absorbers and emitters of infrared radiation.

When a wave is **absorbed** by a substance, the energy it carries is absorbed and makes the substance heat up. It may also create an **alternating current** of the same frequency as the radiation. This principle is used in television and radio aerials, which receive information via radio waves.

Electromagnetic waves obey the wave formula:

Wave speed (metre / second, m/s)	=	**Frequency** (hertz, Hz)	×	**Wavelength** (metre, m)

Visible Light

Light is one type of electromagnetic wave that, together with the other various types of radiation, is in the electromagnetic spectrum.

The seven 'colours of the rainbow' form the **visible spectrum** which, as the name suggests, is the only part of the electromagnetic spectrum that we can see.

The visible spectrum is produced because white light is made up of many different colours. These are refracted by different amounts as they pass through a prism – red light is refracted the least and violet is refracted the most.

Visible light can be used for communication through optical fibres, and for vision and photography.

Electromagnetic wave	Uses	Effects
Radio Waves	• Transmitting radio and TV signals between places – waves with longer wavelengths are reflected by the ionosphere (an electrically charged layer in the atmosphere) so they can send signals between points regardless of the curve of the Earth's surface.	• High levels of exposure for short periods can increase body temperature leading to tissue damage, especially to the eyes.
Microwaves	• Satellite communication networks and mobile phone networks. • Cooking – microwaves are absorbed by water molecules causing them to heat up. • Bluetooth devices.	• May damage or kill cells because they are absorbed by water in the cells, leading to the release of heat. Therefore, care must be taken in the use of microwaves.
Infrared Rays	• Grills, toasters and radiant heaters. • Remote controls for televisions and video recorders. • Optical fibre communication.	• Absorbed by skin and felt as heat. • Excessive amount can cause burns.
Visible Light	See page 98	See page 98
Ultraviolet Rays	• Security coding – a surface coated with special paint absorbs UV and emits visible light. • Sunbathing and sunbeds.	• Passes through skin to the tissues below. Darker skin allows less penetration and provides more protection. • High doses of this radiation can kill cells and low doses can cause skin cancer.
X-Rays	• Producing shadow pictures of bones and metals. • Some cancers can be treated by irradiation with X-rays (radiotherapy).	• Passes through soft tissues, although some is absorbed. • High doses can kill cells and low doses can cause cancer.
Gamma Rays	• Killing cancerous cells. • Killing bacteria on food and surgical instruments.	• Passes through soft tissues – although some is absorbed. • High doses of this radiation can kill cells and low doses can cause cancer.

You need to be able to evaluate the possible hazards associated with the use of different types of electromagnetic radiation.

Example

It's Good to Talk – or is it?

New findings raise concerns that mobile phones could cause cancer and other health problems.

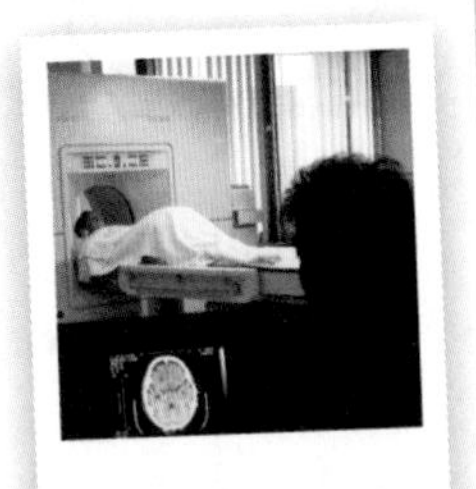

Swedish scientists studying the effects of electromagnetic radiation on red blood cells have found that levels of radiation, equivalent to those emitted by mobile phones, have a significant effect on the attractive forces between cells.

Up until now, the conventional view has been that microwaves could only cause damage at a cellular level if they carried enough energy to break chemical bonds or 'fry' tissue. These new findings might suggest that mobile phones do in fact emit enough energy to affect the bonds.

Experts, however, have been quick to point out that the results were obtained through tests on small groups of cells and provide no real evidence of a danger to health.

There have been other suggestions in the past that mobile phones can cause brain tumours and Alzheimer's disease, but so far research has been inconclusive.

Mobile Phones are Safe

New study has found no link between mobile phones and cancer.

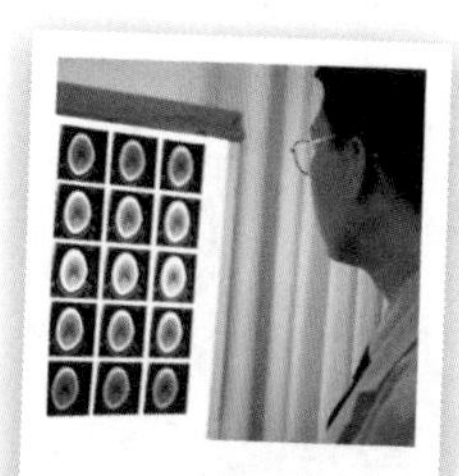

Scientists studying the possible links between mobile phones and brain tumours have reported today that they have found no correlation between using mobile phones and the risk of developing glioma – the most common type of brain tumour.

The study of 2682 people across the UK looked at 966 individuals with diagnosed glioma and 1716 individuals without the condition. It concluded that although a large percentage of the cancer sufferers reported their tumours to be on the side of the head where they held the phone, for regular mobile phone users, there was no increased risk of developing glioma.

Using Mobile Phones

Advantages	Disadvantages
• Easy, convenient method of communication, especially in a vulnerable situation – when car breaks down, alone at night, feel threatened, etc. • Can be used to access the Internet, take pictures, and watch television / video clips. • Easy way to keep in contact when away from home or abroad, e.g. text messages. • Many different tariffs and networks, which makes them affordable. • Can help in solving crime because mobile phones can be tracked.	• Some studies have linked mobile phone use with brain tumours and Alzheimer's disease. • The long-term effects of using mobile phones are not known – studies are still being carried out. • Increasingly, advertising is targeted at younger age groups who would be more vulnerable to any health implications.

Sound Waves

Sounds travel as longitudinal waves. They are produced when something vibrates. The quality of a note depends on the waveform.

Sound cannot travel through a vacuum.

- Sound **reflects** off hard surfaces to produce echoes.
- Sound is **refracted** when it passes into a different medium or substance.
- Sound can be **diffracted** around buildings or land masses, so a person in the 'shadow' of a large building can still hear sounds we might expect to be 'blocked'.

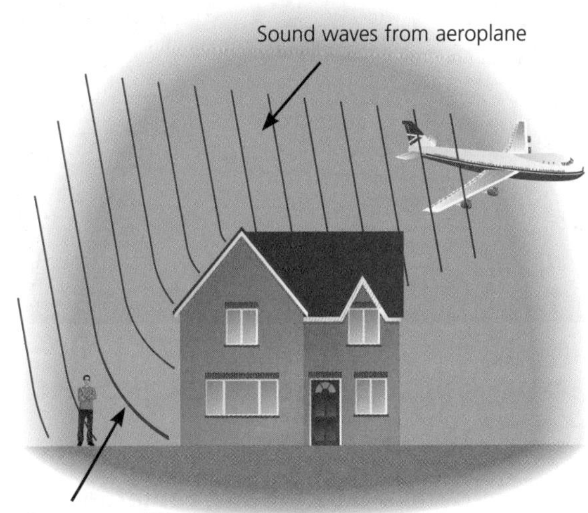

Frequency and Pitch

The frequency of a sound wave is the number of vibrations produced in one second. It is measured in hertz (Hz). Humans can hear sounds in the range of 20–20 000 Hz. The frequency affects the pitch of the sound.

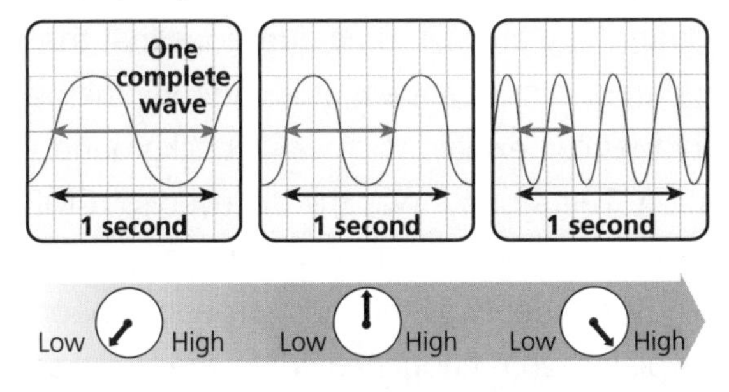

Amplitude and Loudness

Amplitude is the peak of movement of the sound wave from its rest point. The amplitude affects the loudness of the sound.

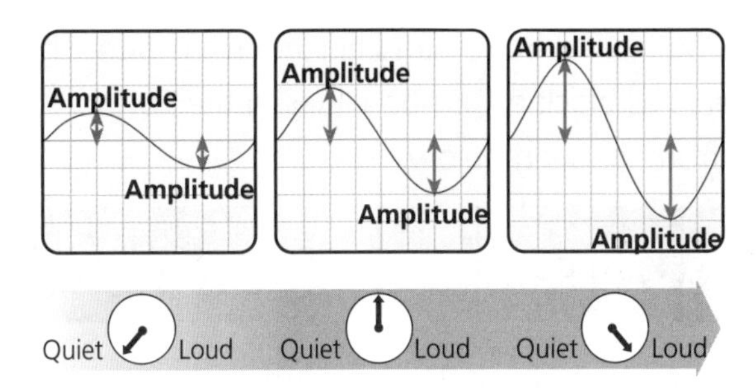

N.B. These diagrams represent the sound wave – remember that sound waves are longitudinal.

The Doppler Effect

When a wave source is moving towards or away from an observer there is a change in the observed wavelength and frequency.

We can detect this change with moving vehicles. For example, when a car approaches you quickly, the sound waves it produces are bunched up, become higher frequency and higher pitched. As the car moves away, the sound waves are stretched out, become lower frequency and lower pitched.

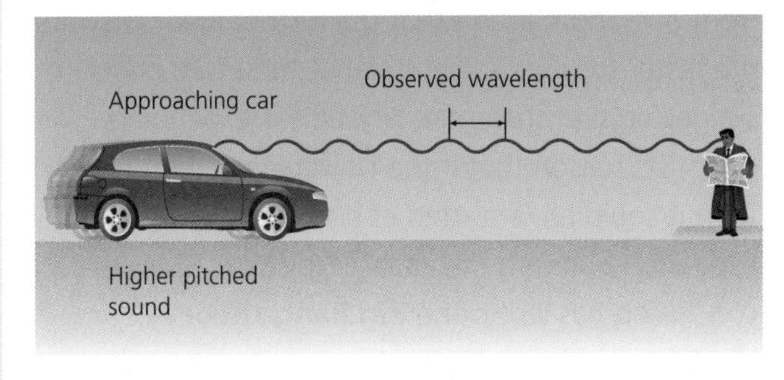

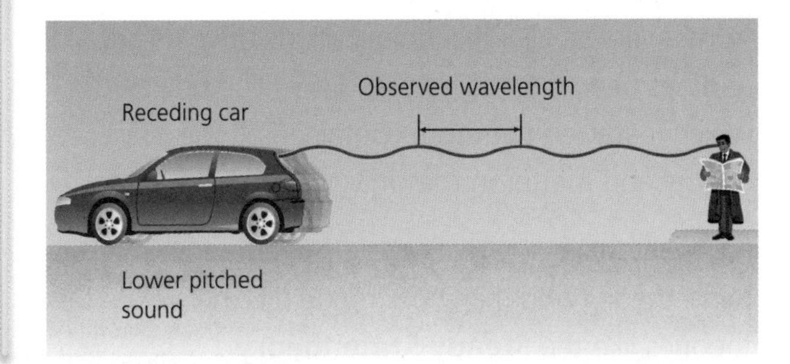

Red-shift

When a light source is moving towards or away from us its frequency appears to change.

When we look at light from the Sun some frequencies of light are not seen because they are absorbed by hydrogen and helium in the Sun. This appears as black absorption lines in the Sun's spectrum, as seen below.

When we look at a galaxy moving away from us the absorption lines have been shifted to lower frequencies, i.e. it has been **red-shifted**, as seen below. If the galaxy was moving towards us its spectrum would be blue-shifted.

Almost all galaxies show red-shift and the further away a galaxy is the more red-shifted it is. This means that the more distant the galaxy the faster it is moving away from us.

The Big Bang Theory

There is evidence that all galaxies are moving away from one another and that the further away they are the faster they are moving. It has therefore been suggested that the entire Universe is expanding. Scientists reason that if the Universe is expanding, then it could have started at a single point and expanded outwards like in an explosion. This conclusion has led to the **Big Bang theory**.

The Big Bang theory states that the Universe started from a very small initial point billions of years ago and has been expanding ever since. The point from which it started therefore contained all the space, energy and matter in the Universe.

For many years scientists could not agree on the Big Bang theory with many preferring alternative theories like the **Steady State theory**.

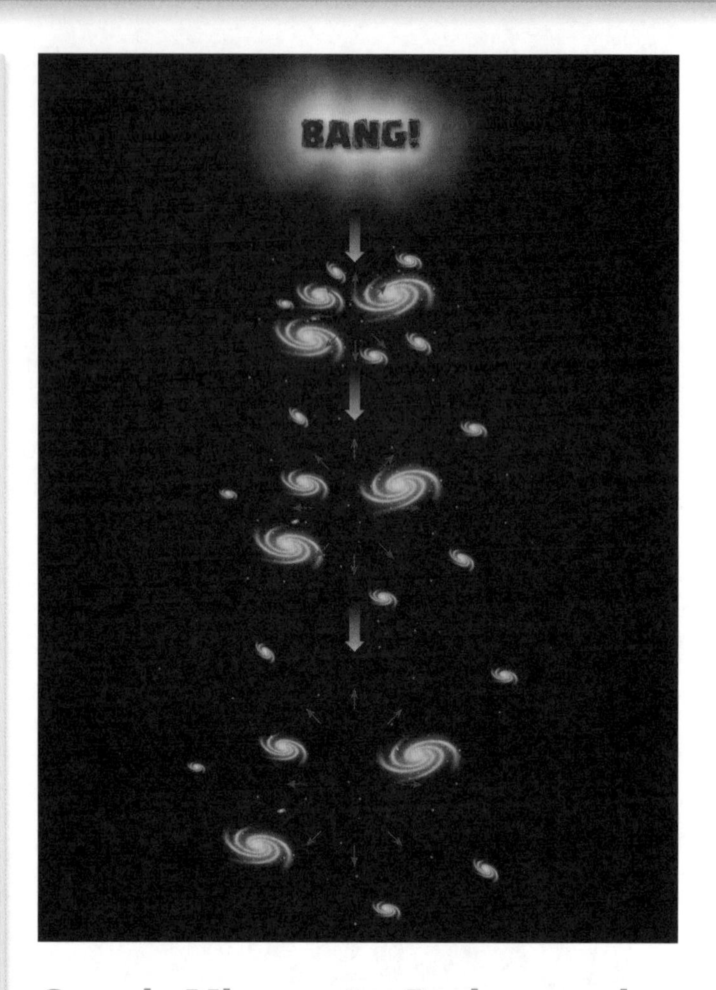

Cosmic Microwave Background Radiation

Cosmic microwave background radiation (CMBR) was discovered in the 1960s. When looking at the space between galaxies with an optical telescope everything appears black. But when using a telescope that is capable of detecting microwaves, a faint glow is seen. This glow is almost the same in all directions and fills the Universe.

It is believed that this background radiation comes from radiation that was present shortly after the beginning of the Universe.

Although there are still some questions that the Big Bang theory cannot answer, it is currently the only theory that can explain microwave background radiation. Therefore, it was the discovery of CMBR that led to the scientific community finally accepting the Big Bang theory as the most probable theory that explains what happened when the Universe began.

1 A wave machine in a 25 m swimming pool generates waves at a frequency of 2 Hz. During operation there are 5 complete waves in the pool.

a) Calculate the speed of the water waves.
Write down the equation you use and then show how you work out your answer. **(4 marks)**

b) What type of waves are water waves? **(1 mark)**

2 The diagram shows the National Grid system.

a) On the diagram label the step-up and step-down transformers. **(2 marks)**

b) When transmitting electricity it is done at a high voltage and low current. Why is this? **(2 marks)**

3 Astronomers have observed that the wavelengths of the light given out from distant galaxies are longer than expected.

a) What is this called? **(1 mark)**

b) Circle the correct option in the following sentence. 'This observation gives evidence for the idea that the Universe is … **(1 mark)**

shrinking **not changing** **expanding**

c) This observation helped lead some astronomers to the Big Bang theory. What other piece of evidence supports the Big Bang theory? **(1 mark)**

4 The diagram shows the energy transfers in a coal-fired power station.

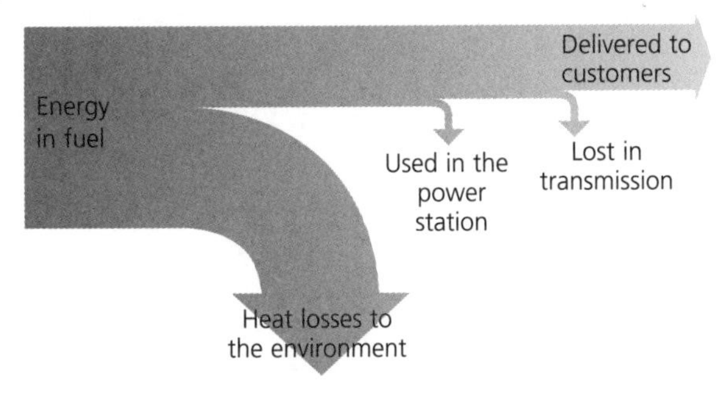

a) What type of energy is stored in the fuel? **(1 mark)**

b) What type of energy is delivered to the customers? **(1 mark)**

c) From the diagram estimate the efficiency of the system and explain how you made your estimate. **(2 marks)**

Answers

Biology
Page 46

1. **a)** A pill not containing a drug i.e. a dummy pill.
   b) These were the control group; To prevent bias in reporting results.
   c) The one with Reduce; The sample size was larger.
   d) Both slimming drugs help people to lose weight; LessU is more effective than Reduce; But even taking a placebo resulted in some weight loss. **[Any two for 2 marks]**
   e) Calorie intake of volunteers; Exercise taken by volunteers; Similar ages; Sex ratio in all groups. **[Any two for 2 marks]**
   f) Tested on animals; Tested on tissue cultures
   g) To check for toxicity; To check efficacy (the drug works); To find optimum dose. **[Any two for 2 marks]**
2. **a)** Hair colour; Shape of their ear lobes.
   b) 23; Nucleus; DNA; Genes
3. **a)** 3 billion years ago.
   b) Unacceptable on religious grounds (God created the world); Not enough evidence; Mechanisms of inheritance not known at the time. **[Any two for 2 marks]**
   c) Competition; Environmental changes; New predators or diseases; Catastrophic event. **[Any two for 2 marks]**

Chemistry
Page 77

1. **a)** 2,8,7
   b) It has 7 electrons in its outermost shell / energy level.

2. **a)** It is a mineral / mixture of minerals / rock from which it is economically viable to extract a metal.
   b) Reduction; Less; Electrolysis

3. **a)** Hydrogen and carbon (only).
   b) C_nH_{2n+2}
   c) i) Heat; A catalyst
   ii) Cracking
   iii) (Short-chain) alkene
   iv) Use bromine water; Bromine water will change from orange to colourless with the alkene but remain orange with the alkane (i.e. bromine is decolourised).

4. **a)** $4Na + O_2 \longrightarrow 2Na_2O$
   b) $H_2SO_4 + 2NaOH \longrightarrow Na_2SO_4 + 2H_2O$

Physics
Page 103

1. **a)** Speed = frequency × wavelength **[1 mark]**
   = 2 × 5 **[1 mark]**
   = 10 **[1 mark]** m/s **[1 mark]**
   b) Transverse
2. **a)** Step-up transformer – label pointing to first box; step-down transformer – label pointing to second box **[1 mark each]**
   b) Wires stay cool; Lose less energy; More efficient **[Any two for 2 marks]**
3. **a)** Red-shift
   b) Expanding
   c) Cosmic microwave background radiation
4. **a)** Chemical
   b) Electrical
   c) 35% (accept 30%–40%) **[1 mark]** ; In total around twice as much is wasted as useful or similar, i.e. only a third gets through to customer **[1 mark]**

Biology

Adaptation – the gradual change of a particular organism over generations to become better suited to its environment

Agar – jelly-like substance made from seaweed that is used to grow bacteria

Auxin – hormone produced by plants

Biomass – the mass of a plant or animal minus the water content

Carcinogen – a substance that causes cancer

Chromosome – a coil of DNA made up of genes, found in the nucleus of plant / animal cells

Clone – an organism that is genetically identical to another organism

Community – the total collection of living organisms within a defined area or habitat

Culture medium – a nutrient system used for the artificial growth of bacteria and other cells

Decay – to rot or decompose

Deficiency disease – a disease caused by the lack of some essential element in the diet

Detritus – organic material formed from dead and decomposing plants and animals

DNA (deoxyribonucleic acid) – nucleic acid molecules that contain genetic information and make up chromosomes

Effector – a part of the body, e.g. a muscle or a gland, that produces a response to a stimulus

Embryo – a ball of cells that will develop into a human / animal baby

Evolve – to change naturally over a period of time

Extinct – a species that has died out

Extremophiles – organisms that can survive extreme environmental conditions, e.g. very high temperatures

Food chain – the feeding relationship between organisms in an ecosystem

Fossil – remains of animals / plants preserved in rocks

FSH (follicle stimulating hormone) – stimulates eggs to mature and produce oestrogen

Gene – part of a chromosome; composed of DNA

Gland – an organ in an animal body used for secreting substances

Gravitropism – directional growth of roots and stems in response to gravity (also called geotropism)

Herbicide – a toxic substance used to destroy unwanted vegetation

Hormone – a regulatory substance that stimulates cells or tissues into action

Immunity – having resistance to a particular disease

Immune system – the body's defence system against infections and diseases (consists of white blood cells and antibodies)

Incubated – grown in a laboratory under controlled conditions

Infectious – a disease that is easily spread, through water, air, etc.

Ions – particles in minerals, such as sodium and potassium that are needed by the body

Leprosy – a contagious bacterial disease affecting the skin and nerves

LH – (luteinising hormone) – stimulates ovulation

Lichen – an organism that is sensitive to pollution

Line transect – line between two points used to investigate the distribution of organisms

Malnourished – suffering from lack of essential food nutrients

Menstrual cycle – the monthly cycle of hormonal changes in a woman

Metabolic rate – the rate at which an animal uses energy over a given time period

Microorganism – a very small organism

MRSA – methicillin-resistant Staphylococcus aureus (or 'superbug'), an antibiotic-resistant bacterium

Mutation – a change in the genetic material of a cell

Neurone – specialised cell that transmits electrical messages or nerve impulses

Obesity – the condition of being very overweight

Oestrogen – a female sex hormone produced in the ovaries

Pathogen – an agent causing a disease

Penicillin – antibiotic drug used to treat bacterial infection (discovered by Alexander Fleming)

Pesticide – a substance used for destroying insects or other pests

Petri dish – a round, shallow dish used to grow bacteria

Phototropism – the production of auxin on the shaded side of a plant shoot, which causes it to grow towards the Sun

Pituitary – a small gland at the base of the brain that produces hormones

Placebo – a substance given to patients in drug trials; does not contain a drug

Predator – an animal that hunts, kills and eats other animals

Receptor – a sense organ; e.g. eyes, ears, nose, etc.

Recreation – for pleasure

Reflex action – an involuntary action, e.g. removing hand from a hot plate

Saturated fats – animal fat, considered to be unhealthy

Sterilised – free from all microorganisms

Surface area – the external area of a living thing

Synapse – the gap between two nerve cells

Toxin – a poison produced by a living organism

Vaccine – a liquid preparation used to make the body produce antibodies in order to provide protection against disease

Variation – differences between individuals of the same species

Chemistry

Alkane – a saturated hydrocarbon (C_nH_{2n+2})

Alkene – an unsaturated hydrocarbon containing at least one carbon–carbon double bond (C_nH_{2n})

Alloy – a mixture of two or more metals or a mixture of one metal and a non-metal

Atom – the smallest part of an element that can enter into chemical reactions

Atomic number – the number of protons in an atom

Biofuel – a fuel produced from plant material

Catalyst – a substance that increases the rate of a chemical reaction while remaining chemically unchanged itself

Chemical formula – a way of showing the elements present in a substance

Chemical reaction – a process in which one or more substances are changed into others

Compound – a substance consisting of two or more different elements that are chemically combined

Decompose – to break down

Distillation – the process of boiling a liquid and condensing its vapours

Electrolysis – the process by which an electric current causes a solution to undergo chemical decomposition into its elements

Electron – a negatively charged particle orbiting the nucleus of an atom. In a neutral atom the number of electrons equals the number of protons

Element – a substance that consists of only one type of atom

Emulsion – a liquid dispersed in another liquid, e.g. oil and water mixed together

Fossil fuel – a fuel formed in the ground over millions of years from the remains of dead plants and animals; a non-renewable fuel

Fuel – a substance that releases heat or energy when combining with oxygen

Hydrocarbon – a compound containing only hydrogen and carbon

Hydrogenation – the process in which hydrogen is used to harden vegetable oils

Mass number – the number of protons plus the number of neutrons in an atom

Mixture – two or more different substances that are not chemically combined

Molecule – the simplest structural unit of an element or covalent compound

Monomer – the small molecules that combine to make up a polymer

Neutron – a particle in the nucleus of an atom; has no electrical charge

Non-biodegradable – a substance that does not decompose naturally

Nucleus – the small central core of an atom consisting of protons and neutrons

Ore – a naturally occurring mineral from which a metal can be extracted (if economically viable)

Polymer – a giant long-chain molecule made by combining lots of monomers

Polymerisation – the reaction in which a polymer is formed by joining together lots of monomers

Proton – a positively charged particle in the nucleus of an atom

Transition elements – a block of metallic elements in the middle of the Periodic Table

Physics

Alternating current – an electric current that changes direction of flow continuously

Amplitude – the height from the middle of a wave to the top of a crest or bottom of a trough

Big Bang theory – the theory that the Universe started with a big explosion

Compressions – the points in a longitudinal wave where the particles are squashed together

Cosmic microwave background radiation – microwave radiation thought to be evidence of the Big Bang

Condensation – the change of state from a gas to a liquid

Conductor – a substance that readily transfers heat or energy

Convection – the transfer of heat energy in liquids and gases

Current – the rate of flow of electrical charge through a conductor; measured in amperes (A)

Diffraction – the spreading out of waves that have passed through a narrow gap

Doppler effect – the apparent change of frequency for a moving wave source

Efficiency – the ratio of energy output to energy input, expressed as a percentage

Electromagnetic spectrum – a continuous spectrum of transverse waves that can travel through a vacuum

Energy – the capacity of a physical system to do work; measured in joules (J)

Frequency – the number of waves per second

Ion – a charged particle formed when an atom gains or loses electrons

Kilowatt – a unit for measuring power, equal to 1000 watts

Kilowatt-hour – the amount of electrical energy used by a 1 kilowatt device in 1 hour

Longitudinal – waves in which the vibrations are parallel to the direction of propagation of energy

National Grid – the system of power stations, transformers and pylons that transmit electricity across the country

Non-renewable – energy sources that cannot be replaced in a lifetime

Pitch – how high or low a sound is, changes with frequency

Plane mirror – a flat (non-curved) mirror

Power – the rate of doing work; measured in watts (W)

Rarefactions – the points in a longitudinal wave where the particles are further apart

Real image – an image produced by rays of light meeting at a point (can be projected on a screen)

Red-shift – the shift in observed spectra of light emitted from distant galaxies because they are moving away

Reflection – a wave (e.g. light or sound) that is thrown back from a surface

Refraction – the change in direction of a wave as it passes from one medium to another

Renewable – energy sources that can be replaced as they are used

Specific heat capacity – the energy needed to heat 1kg of a substance by 1°C

Steady State theory – the theory that the Universe has always existed

Thermal energy – heat energy

Thermal radiation – the transfer of heat energy as infrared waves

Transfer – to move energy from one place to another

Transformer – an electrical device used to change the voltage of alternating currents

Transmission – the sending of information or electricity over a communications line or a circuit

Transverse – waves in which the vibrations are perpendicular to the direction of propagation energy

Virtual image – an image produced where rays of light only appear to meet

Visible spectrum – the part of the electromagnetic spectrum detectable by the human eye

Voltage/potential difference (p.d.) – the energy transferred by electric charge, per unit charge; expressed in volts (V)

Wavelength – the height of the wave from the middle to the top

Notes

Electronic Structure of the First 20 Elements

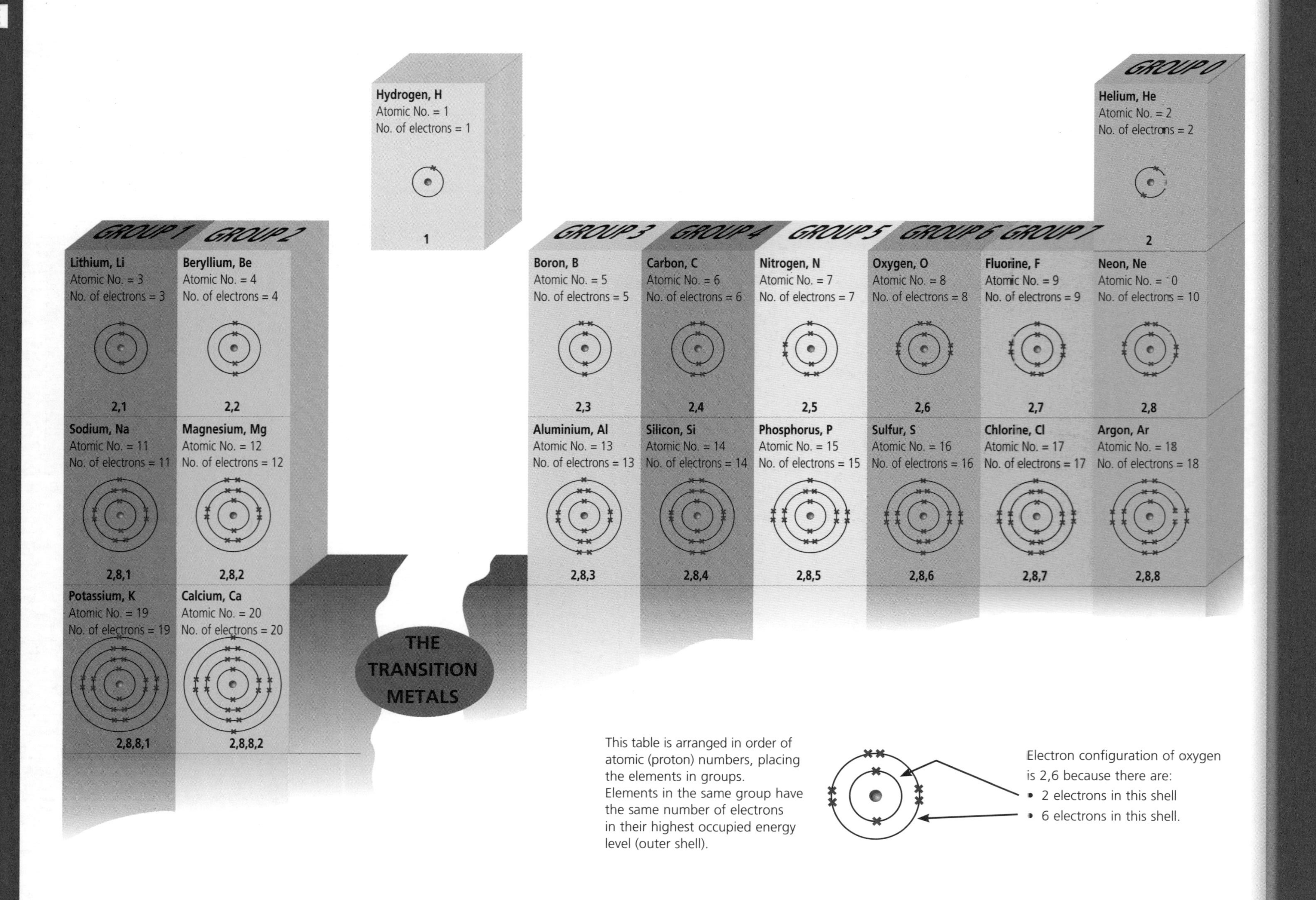

The Modern Periodic Table

Key

relative atomic mass **atomic symbol** name atomic (proton) number

1 **H** hydrogen 1

1	2											3	4	5	6	7	0
																	4 **He** helium 2
7 **Li** lithium 3	9 **Be** beryllium 4											11 **B** boron 5	12 **C** carbon 6	14 **N** nitrogen 7	16 **O** oxygen 8	19 **F** fluorine 9	20 **Ne** neon 10
23 **Na** sodium 11	24 **Mg** magnesium 12											27 **Al** aluminium 13	28 **Si** silicon 14	31 **P** phosphorus 15	32 **S** sulfur 16	35.5 **Cl** chlorine 17	40 **Ar** argon 18
39 **K** potassium 19	40 **Ca** calcium 20	45 **Sc** scandium 21	48 **Ti** titanium 22	51 **V** vanadium 23	52 **Cr** chromium 24	55 **Mn** manganese 25	56 **Fe** iron 26	59 **Co** cobalt 27	59 **Ni** nickel 28	63.5 **Cu** copper 29	65 **Zn** zinc 30	70 **Ga** gallium 31	73 **Ge** germanium 32	75 **As** arsenic 33	79 **Se** selenium 34	80 **Br** bromine 35	84 **Kr** krypton 36
85 **Rb** rubidium 37	88 **Sr** strontium 38	89 **Y** yttrium 39	91 **Zr** zirconium 40	93 **Nb** niobium 41	96 **Mo** molybdenum 42	[98] **Tc** technetium 43	101 **Ru** ruthenium 44	103 **Rh** rhodium 45	106 **Pd** palladium 46	108 **Ag** silver 47	112 **Cd** cadmium 48	115 **In** indium 49	119 **Sn** tin 50	122 **Sb** antimony 51	128 **Te** tellurium 52	127 **I** iodine 53	131 **Xe** xenon 54
133 **Cs** caesium 55	137 **Ba** barium 56	139 **La*** lanthanum 57	178 **Hf** hafnium 72	181 **Ta** tantalum 73	184 **W** tungsten 74	186 **Re** rhenium 75	190 **Os** osmium 76	192 **Ir** iridium 77	195 **Pt** platinum 78	197 **Au** gold 79	201 **Hg** mercury 80	204 **Tl** thallium 81	207 **Pb** lead 82	209 **Bi** bismuth 83	[209] **Po** polonium 84	[210] **At** astatine 85	[222] **Rn** radon 86
[223] **Fr** francium 87	[226] **Ra** radium 88	[227] **Ac*** actinium 89	[261] **Rf** rutherfordium 104	[262] **Db** dubnium 105	[266] **Sg** seaborgium 106	[264] **Bh** bohrium 107	[277] **Hs** hassium 108	[268] **Mt** meitnerium 109	[271] **Ds** darmstadtium 110	[272] **Rg** roentgenium 111	Elements with atomic numbers 112–116 have been reported but not fully authenticated						

*The Lanthanides (atomic numbers 58–71) and the Actinides (atomic numbers 90–103) have been omitted.

Cu and **Cl** have not been rounded to the nearest whole number.

Index